Information Circular 9517

Best Practices for Dust Control in Coal Mining

By Jay F. Colinet, James P. Rider, Jeffrey M. Listak, John A. Organiscak, and Anita L. Wolfe

DEPARTMENT OF HEALTH AND HUMAN SERVICES
Centers for Disease Control and Prevention
National Institute for Occupational Safety and Health
Office of Mine Safety and Health Research
Pittsburgh, PA • Spokane, WA

January 2010

This document is in the public domain and may be freely copied or reprinted.

Disclaimer

Mention of any company or product does not constitute endorsement by the National Institute for Occupational Safety and Health (NIOSH). In addition, citations to Web sites external to NIOSH do not constitute NIOSH endorsement of the sponsoring organizations or their programs or products. Furthermore, NIOSH is not responsible for the content of these Web sites. All Web addresses referenced in this document were accessible as of the publication date.

Ordering Information

To receive documents or other information about occupational safety and health topics, contact NIOSH at

> Telephone: **1–800–CDC–INFO** (1–800–232–4636)
> TTY: 1–888–232–6348
> e-mail: cdcinfo@cdc.gov
>
> or visit the NIOSH Web site at **www.cdc.gov/niosh**.

For a monthly update on news at NIOSH, subscribe to NIOSH *eNews* by visiting **www.cdc.gov/niosh/eNews**.

DHHS (NIOSH) Publication No. 2010–110

January 2010

SAFER • HEALTHIER • PEOPLE™

CONTENTS

Page

Introduction 1
Chapter 1.—Health effects of overexposure to respirable coal and silica dust 3
Chapter 2.—Sampling to quantify respirable dust generation 11
Chapter 3.—Controlling respirable dust on longwall mining operations 17
Chapter 4.—Controlling respirable dust on continuous mining operations 41
Chapter 5.—Controlling respirable silica dust at surface mines 65

ILLUSTRATIONS

1-1. Normal lung and a lung from a miner diagnosed with CWP 4
1-2. Trends in CWP prevalence among examinees employed at underground coal mines by years of experience (tenure) 5
1-3. Section of a freeze-dried human lung with silicosis 7
1-4. Percentage of MSHA inspector samples during 2003–2007 that exceeded reduced PELs 7
2-1. Gravimetric sampling pump, cyclone, and filter cassette 12
2-2. Example of dust measurements obtained with the pDR 13
2-3. PDM with TEOM removed 13
2-4. Sampling locations used to isolate dust generated by a continuous miner 14
2-5. Mobile sampling used to quantify shearer dust 15
2-6. Sampling locations around a surface drill 16
3-1. Rotary brush cleans the conveying side of the belt 20
3-2. Water sprays and belt wiper used to reduce dust from the nonconveying side of the belt as its returns 20
3-3. Enclosed stageloader/crusher and location of water sprays 22
3-4. High-pressure water scrubber installed on top of crusher 23
3-5. Gob curtain increases airflow down the face 24
3-6. Ventilation patterns around shearer without and with a cutout curtain 25
3-7. Shearer-clearer directional spray system 27
3-8. Venturi sprays mounted on headgate splitter arm 28
3-9. Headgate splitter arm with flat-fan sprays mounted on gob side of belting 28
3-10. Directional sprays mounted on face side of shearer body 29
3-11. Position of splitter arm may allow dust to migrate into walkway 30
3-12. Raised deflector plate can enhance the effectiveness of the directional spray system 31
3-13. Crescent sprays located on shearer ranging arm 32
3-14. Spray manifold mounted on tailgate end of shearer body 33
3-15. Shield sprays located on the underside of the canopy 34
3-16. Schematic of ventilated shearer drum 35
4-1. Spray types used for dust control in mining 42
4-2. Relative spray effectiveness of four spray nozzles used in mining 44
4-3. Spray location impact on dust rollback 45
4-4. Antirollback spray system for miner 46
4-5. Air-moving effectiveness of different spray types 47
4-6. Components and design of a flooded-bed scrubber 49
4-7. Cleaning scrubber filter panel with water spray 50

CONTENTS—Continued

Page

4-8. Cleaning the demister with a water nozzle 50
4-9. Dust collection efficiency of scrubber filter panels 51
4-10. Proper bit design can lower dust generation 53
4-11. Modified cutting cycle can lower dust generation 53
4-12. Schematic of a blowing ventilation system 54
4-13. Schematic of an exhaust ventilation system 56
4-14. Dust collector box with collector bag installed 58
4-15. Schematic of roof bolter dust collector components 58
4-16. Prototype of canopy curtain 60
5-1. Typical dry dust collection system used on surface drills 66
5-2. Water separator discharging water before it reaches the drill bit 68
5-3. Increase in dust when a wet haul road dries 71
5-4. Staging curtains used to prevent dust from billowing out of enclosure 73
5-5. Tire-stop water spray system reduces dust rollback under the dumping vehicle 74

TABLES

5-1. Respirable dust sampling results of enclosed cab field studies 69

ACRONYMS AND ABBREVIATIONS USED IN THIS REPORT

CWHSP	Coal Workers' Health Surveillance Program
CWP	Coal workers' pneumoconiosis
DO	designated occupation
HVAC	heating, ventilation, and air conditioning
IARC	International Agency for Research on Cancer
ILO	International Labour Office
MSHA	Mine Safety and Health Administration
NIOSH	National Institute for Occupational Safety and Health
PDM	personal dust monitor
pDR	personal DataRAM
PEL	permissible exposure limit
PMF	progressive massive fibrosis
TEOM	tapered-element oscillating microbalance

UNIT OF MEASURE ABBREVIATIONS USED IN THIS REPORT

cfm	cubic foot per minute
cm	centimeter
fpm	foot per minute
ft	foot
ft/min	foot per minute
gpm	gallon per minute
hr	hour
in	inch
in w.g.	inch water gauge
kPa	kilopascal
lpm	liter per minute
m/sec	meter per second
mg/m^3	milligram per cubic meter
mm	millimeter
mph	miles per hour
$\mu g/m^3$	microgram per cubic meter
psi	pound-force per square inch
sec	second

BEST PRACTICES FOR DUST CONTROL IN COAL MINING

By Jay F. Colinet,[1] James P. Rider,[2] Jeffrey M. Listak,[3] John A. Organiscak,[3] and Anita L. Wolfe[4]

INTRODUCTION

Respirable dust exposure has long been known to be a serious health threat to workers in many industries. In coal mining, overexposure to respirable coal mine dust can lead to coal workers' pneumoconiosis (CWP). CWP is a lung disease that can be disabling and fatal in its most severe form. In addition, miners can be exposed to high levels of respirable silica dust, which can cause silicosis, another disabling and/or fatal lung disease. Once contracted, there is no cure for CWP or silicosis. The goal, therefore, is to limit worker exposure to respirable dust to prevent development of these diseases.

The passage of the Federal Coal Mine Health and Safety Act of 1969 established respirable dust exposure limits, dust sampling requirements for inspectors and mine operators, a voluntary x-ray surveillance program to identify CWP in underground coal miners, and a benefits program to provide compensation to affected workers and their families. The tremendous human and financial costs resulting from CWP and silicosis in the U.S. underground coal mine workforce are shown by the following statistics:

- During 1970–2004, CWP was a direct or contributing cause of 69,377 deaths of U.S. underground coal mine workers.
- During 1980–2005, over $39 billion in CWP benefits were paid to underground coal miners and their families.
- Recent x-ray surveillance data for 2000–2006 show an increase in CWP cases. Nearly 8% of examined underground coal miners with 25 or more years of experience were diagnosed with CWP.
- "Continuous miner operator" is the most frequently listed occupation on death certificates that record silicosis as the cause of death.

In light of the ongoing severity of these lung diseases in coal mining, this handbook was developed to identify available engineering controls that can help the industry reduce worker exposure to respirable coal and silica dust. The controls discussed in this handbook range from long-utilized controls that have developed into industry standards to newer controls that are still being optimized. The intent was to identify the best practices that are available to control respirable dust levels in underground and surface coal mining operations. This handbook

---

[1]Supervisory mining engineer, Office of Mine Safety and Health Research, National Institute for Occupational Safety and Health, Pittsburgh, PA.
[2]Operations research analyst, Office of Mine Safety and Health Research, National Institute for Occupational Safety and Health, Pittsburgh, PA.
[3]Mining engineer, Office of Mine Safety and Health Research, National Institute for Occupational Safety and Health, Pittsburgh, PA.
[4]Public health advisor, Division of Respiratory Disease Studies, National Institute for Occupational Safety and Health, Morgantown, WV.

provides general information on the control technologies along with extensive references. In some cases, the full reference(s) will need to be consulted to gain in-depth information on the testing or implementation of the control of interest.

The handbook is divided into five chapters. Chapter 1 discusses the health effects of exposure to respirable coal and silica dust. Chapter 2 discusses dust sampling instruments and sampling methods. Chapters 3, 4, and 5 focus on dust control technologies for longwall mining, continuous mining, and surface mining, respectively.

Finally, it must be stressed that after control technologies are implemented, the ultimate success of ongoing protection for workers depends on continued maintenance of these controls. NIOSH researchers have often seen appropriate controls installed, but worker overexposures occurred because of the lack of proper maintenance of these controls.

CHAPTER 1.—HEALTH EFFECTS OF OVEREXPOSURE TO RESPIRABLE COAL AND SILICA DUST

By Anita L. Wolfe[1] and Jay F. Colinet[2]

Pneumoconioses are lung diseases caused by the inhalation and deposition of mineral dusts in the lungs. Pneumoconioses associated with working in a high-risk, mineral-related industry such as mining are coal workers' pneumoconiosis (CWP) and silicosis. Once contracted, these diseases cannot be cured. Therefore, it is critical to limit worker exposure to airborne respirable dust to prevent these diseases.

COAL WORKERS' PNEUMOCONIOSIS (CWP)

CWP, commonly called black lung disease, is a chronic lung disease that results from the inhalation and deposition of coal dust in the lung and the lung tissue's reaction to its presence. It most often affects those who mine, process, or ship coal. In addition to CWP, coal mine dust exposure increases a miner's risk of developing chronic bronchitis, chronic obstructive pulmonary disease, and pathologic emphysema.

With continued exposure to the dust, the lungs undergo structural changes that are eventually seen on a chest x-ray. In the simple stages of disease (simple CWP), there may be no symptoms. However, when symptoms do develop, they include cough (with or without mucus), wheezing, and shortness of breath (especially during exercise). Figure 1-1 shows a normal lung and a lung from a miner who has been diagnosed with CWP. In the more advanced stages of disease, the structural changes in the lung are called fibrosis. Progressive massive fibrosis (PMF) is the formation of tough, fibrous tissue deposits in the areas of the lung that have become irritated and inflamed. With PMF the lungs become stiff and their ability to expand fully is reduced. This ultimately interferes with the lung's normal exchange of oxygen and carbon dioxide, and breathing becomes very difficult. The patient's lips and fingernails may have a bluish tinge, and there may be fluid retention and signs of heart failure. If a person has inhaled too much coal dust, simple CWP can progress to PMF.

Simple CWP is characterized by the presence of small opacities (opaque spots) on the chest x-ray that are less than 10 mm in diameter. The profusion (density) of small opacities is classified as major category 1, 2, or 3 as defined by the International Labour Office (ILO) guidelines [ILO 1980]. Category 0 is defined as the absence of small opacities or opacities that are less profuse than the lower limit of category 1. Within the 12-point ILO profusion scale, each major category may be followed by a subcategory, if an adjacent main category was considered during classification (e.g., classification 1/2 was judged as category 1, but category 2 was seriously considered) [NIOSH 1995].

---

[1]Public health advisor, Division of Respiratory Disease Studies, National Institute for Occupational Safety and Health, Morgantown, WV.
[2]Supervisory mining engineer, Office of Mine Safety and Health Research, National Institute for Occupational Safety and Health, Pittsburgh, PA.

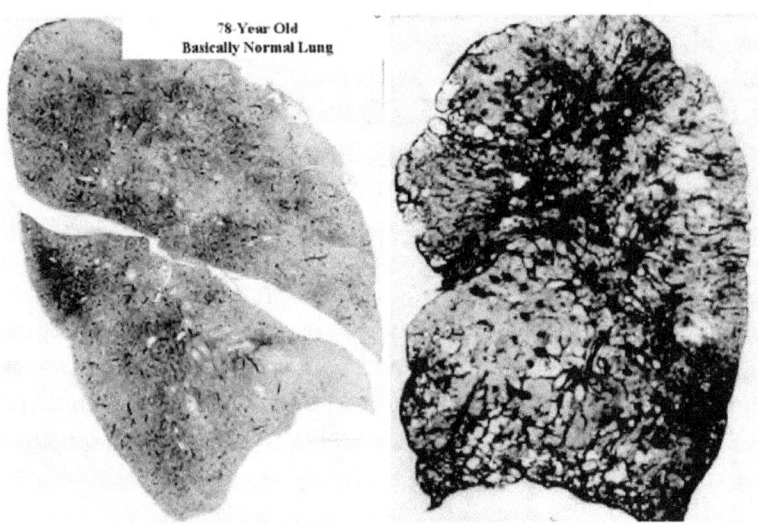

Figure 1-1.—Normal lung *(left)* and a lung from a miner diagnosed with CWP *(right)*.

PMF is classified radiographically as category A, B, or C when large opacities with a combined area of 1 cm or larger are found on the chest x-ray. PMF usually develops in miners already affected by simple CWP, but can develop in miners with no previous radiographic evidence of simple CWP [NIOSH 1995].

There is no specific therapy for these diseases. Primary prevention of lung disease in miners must include continued efforts to reduce coal mine dust exposure. Medical management is best directed at prevention, early recognition, and treatment of complications. The major clinical challenges are the recognition and management of airflow obstruction, respiratory infection, hypoxemia (an abnormally low amount of oxygen in the blood), respiratory failure, cor pulmonale (enlargement of the right side of the heart), arrhythmias (abnormal heart rhythm), and pneumothorax (collapsed lung).

Since passage of the Federal Coal Mine Health and Safety Act of 1969, the Mine Safety and Health Administration (MSHA) enforces regulations designed to limit mine workers' exposure to respirable coal mine dust to 2 mg/m^3 or less if the silica content in the sample is less than 5%. Periodic sampling is conducted by MSHA inspectors and mine operators to demonstrate compliance with this dust limit. In underground coal mines, airborne dust concentrations are typically the highest for workers involved in the extraction of coal at the mining face. Longwall shearer operators, jack setters, and continuous miner operators are occupations with greater potential for exposure to excessive levels of respirable coal mine dust. Workers in some aboveground coal mining operations also have increased exposure to coal mine dust. These include workers at preparation plants where crushing, sizing, washing, and blending of coal are performed and at tipples where coal is loaded into trucks, railroad cars, river barges, or ships.

Also included in the 1969 Act was the establishment of the NIOSH Coal Workers' Health Surveillance Program (CWHSP). As part of this program, underground coal miners are periodically offered the opportunity to voluntarily receive a chest x-ray (free of charge to the miner) in an effort to detect the presence of CWP. The rates of black lung steadily declined through 1999. However, recent data from NIOSH [2008] show that the declines have stopped and rates are actually starting to rise (see Figure 1-2). For miners with 25 or more years of experience who were examined in the CWHSP after the year 2000, the rate of black lung being found has nearly doubled. In addition, disease is showing up in younger miners, and miners are progressing from the beginning stages of black lung disease to the more advanced PMF at a faster rate. In 2004, the deaths of 703 miners were attributed to CWP. (For additional statistics, see: NIOSH [2008]).

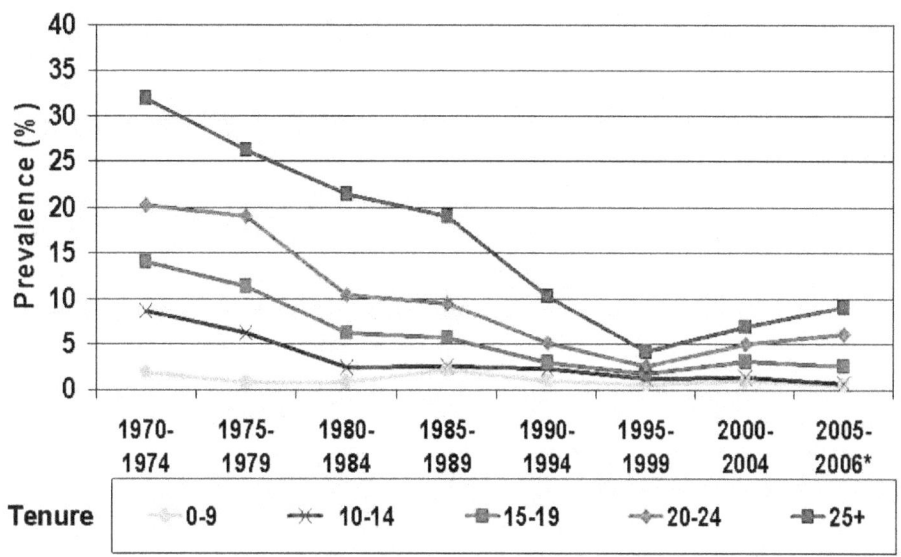

Figure 1-2.—Trends in CWP prevalence among examinees employed at underground coal mines by years of experience (tenure). (Source: NIOSH Coal Workers' X-ray Surveillance Program).

30 CFR[3] 90 establishes procedures for miners who have developed evidence of pneumoconiosis to work in an area of a mine where the average concentration of respirable dust in the mine atmosphere during each shift is continuously maintained at or below 1.0 mg/m^3. The rule sets forth procedures for miners to exercise this option and establishes the right of miners to retain their regular rate of pay and receive wage increases. The rule also sets forth the mine operator's obligations, including respirable dust sampling requirements for Part 90 miners. The goal is to prevent further development of the pneumoconiosis in the affected miner.

---

[3]Code of Federal Regulations. See CFR in references.

SILICOSIS

Occupational exposures to respirable crystalline silica occur in a variety of industries and occupations because of its extremely common natural occurrence. Workers with high exposure to crystalline silica include miners, sandblasters, tunnel builders, silica millers, quarry workers, foundry workers, and ceramics and glass workers. Silica refers to the chemical compound silicon dioxide (SiO_2), which occurs in a crystalline or noncrystalline (amorphous) form [NIOSH 2002]. Crystalline silica may be found in more than one form: alpha quartz, beta quartz, tridymite, and cristobalite [Ampian and Virta 1992; Heaney 1994]. In nature, the alpha form of quartz is the most common [Virta 1993]. This form is so abundant that the term "quartz" is often used instead of the general term "crystalline silica" [USBM 1992; Virta 1993].

Quartz is a common component of rocks. Mine workers are potentially exposed to quartz dust when rock within or adjacent to the coal seams is cut, crushed, and transported. Occupational exposures to respirable crystalline silica are associated with the development of silicosis, lung cancer, pulmonary tuberculosis, and airways diseases. These exposures may also be related to the development of autoimmune disorders, chronic renal disease, and other adverse health effects. In 1996, the International Agency for Research on Cancer reviewed the published experimental and epidemiologic studies of cancer in animals and workers exposed to respirable crystalline silica. The IARC concluded that there was sufficient evidence to classify silica as a human carcinogen [IARC 1997].

Silicosis is also a fibrosing disease of the lungs caused by the inhalation, retention, and pulmonary reaction to the crystalline silica. The main symptom of silicosis is usually dyspnea (difficult or labored breathing and/or shortness of breath). This is first noted with activity or exercise and later as the functional reserve of the lung is also lost at rest. However, in the absence of other respiratory disease, there may be no shortness of breath and the disease may first be detected through an abnormal chest x-ray. The x-ray may at times show quite advanced disease with only minimal symptoms. The appearance or progression of dyspnea may indicate other complications, including tuberculosis, airways obstruction, PMF, or cor pulmonale. A productive cough is often present.

A worker may develop one of three types of silicosis, depending on the airborne concentrations of respirable crystalline silica that were inhaled:

(1) *Chronic Silicosis:* Usually occurs after 10 or more years of exposure at relatively low concentrations. Swellings caused by the silica dust form in the lungs and chest lymph nodes. This disease may cause people to have trouble breathing and may be similar to chronic obstructive pulmonary disease.

(2) *Accelerated Silicosis:* Develops 5–10 years after the first exposure. Swelling in the lungs and symptoms occur faster than in chronic silicosis.

(3) *Acute Silicosis:* Develops after exposure to high concentrations of respirable crystalline silica and results in symptoms within a period of a few weeks to 5 years after initial exposure [Parker and Wagner 1998; Peters 1986]. The lungs become very inflamed and can fill with fluid, causing severe shortness of breath and low blood oxygen levels.

PMF can occur in either simple or accelerated silicosis, but is more common in the latter. Figure 1-3 shows a lung that has been damaged by silicosis.

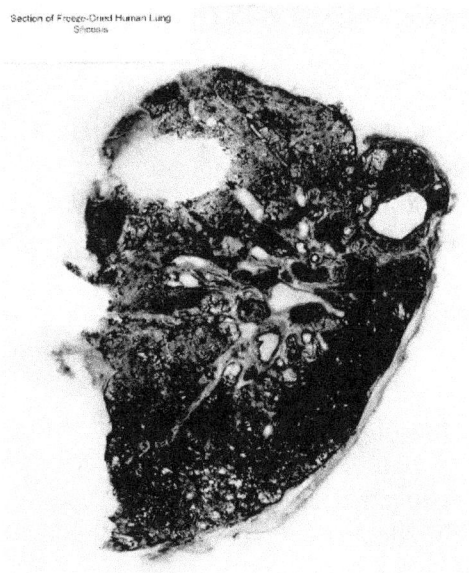

Figure 1-3.—Section of a freeze-dried human lung with silicosis.

In an effort to prevent the development of silicosis, MSHA regulates the exposure of mine workers to silica. For coal mining operations, quartz levels up to 5% in compliance dust samples do not alter the respirable dust standard of 2 mg/m^3. However, if the percent of quartz in the sample exceeds 5%, a reduced dust standard is calculated by dividing 10 by the percent quartz. For example, if a sample contains 10% quartz, the reduced standard would be equal to 1 mg/m^3 (10 ÷ 10% quartz). In essence, these regulations limit the exposure to respirable quartz to 100 µg/m^3, although this limit is not specifically quantified in the regulations.

MSHA compliance sampling data identify those occupations in coal mining that are at high risk for overexposure to quartz. Figure 1-4 shows the percentage of samples collected by MSHA inspectors that exceeded reduced permissible exposure limits (PELs) for several high-risk occupations in coal mining.

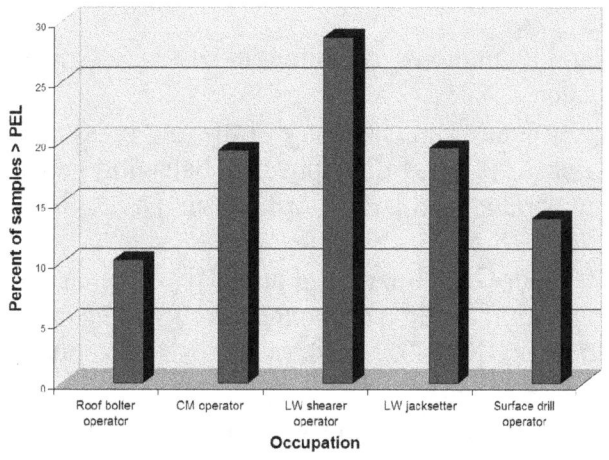

Figure 1-4.—Percentage of MSHA inspector samples during 2003–2007 that exceeded reduced PELs.

DIAGNOSIS AND TREATMENT OF PNEUMOCONIOSES

A doctor may diagnose CWP or silicosis based on the combination of an appropriate history of exposure to coal mine dust or silica, compatible changes in chest imaging or lung pathology, and absence of plausible alternative diagnoses. A chest radiograph is often sufficient for diagnosis, but in some cases a computed tomography (CT) scan of the chest can be helpful. Lung biopsy, a procedure where a sample of lung tissue is taken for lab examination, is not usually required if a compatible exposure history and findings on chest imaging are present. Pulmonary function tests and blood tests to measure the amounts of oxygen and carbon dioxide in the blood (arterial blood gases) can help in objectively assessing the level of impairment caused by CWP or silicosis.

Epidemiologic studies of gold miners in South Africa, granite quarry workers in Hong Kong, metal miners in Colorado, and coal miners in Scotland have shown that chronic silicosis may develop or progress even after occupational exposure to silica has been discontinued [Hessel et al. 1988; Hnizdo and Sluis-Cremer 1993; Ng et al. 1987; Kreiss and Zhen 1996; Miller et al. 1998]. Therefore, removing a worker from exposure after diagnosis does not guarantee that silicosis or silica-related disease will stop progressing or that an impaired worker's condition will stabilize.

Treatment of CWP or silicosis may include use of bronchodilators (medications to open the airways) or supplemental oxygen use. Once disease is detected, it is important to protect the lungs against respiratory infections. Thus, a doctor may recommend vaccinations to prevent influenza and pneumonia. In some cases of severe disease, a lung transplant may be recommended. Prognosis depends on the specific type of pneumoconiosis and the duration and level of dust exposure.

There is no cure for these lung diseases, and they cannot be reversed. Effective control technologies must be implemented and continually maintained to prevent the development of the disease.

REFERENCES

Ampian SG, Virta RL [1992]. Crystalline silica overview: occurrence and analysis. Washington, DC: U.S. Department of the Interior, Bureau of Mines, IC 9317. NTIS No. PB92-200997.

CFR. Code of federal regulations. Washington, DC: U.S. Government Printing Office, Office of the Federal Register.

Heaney PJ [1994]. Structure and chemistry of the low-pressure silica polymorphs. In: Heaney PJ, Prewitt CT, Gibbs GV, eds. Silica: physical behavior, geochemistry, and materials applications. Reviews in mineralogy. Vol. 29. Washington, DC: Mineralogical Society of America.

Hessel PA, Sluis-Cremer GK, Hnizdo E, Faure MH, Thomas RG, Wiles FJ [1988]. Progression of silicosis in relation to silica dust exposure. Ann Occup Hyg 32(Suppl 1):689–696.

Hnizdo E, Sluis-Cremer GK [1993]. Risk of silicosis in a cohort of white South African gold miners. Am J Ind Med 24:447–457.

IARC [1997]. IARC monographs on the evaluation of carcinogenic risks to humans: Silica, some silicates, coal dust and para-aramid fibrils. Vol. 68. Lyon, France: World Health Organization, International Agency for Research on Cancer.

ILO [1980]. Guidelines for the use of ILO international classification of radiographs of pneumoconiosis. Rev. ed. Occupational Safety and Health Series No. 22. Geneva, Switzerland: International Labour Office.

Kreiss K, Zhen B [1996]. Risk of silicosis in a Colorado mining community. Am J Ind Med 30:529–539.

Miller BG, Hagen S, Love RG, Soutar CA, Cowie HA, Kidd MW, Robertson A [1998]. Risks of silicosis in coalworkers exposed to unusual concentrations of respirable quartz. Occup Environ Med 55:52–58.

Ng TP, Chan SL, Lam KP [1987]. Radiological progression and lung function in silicosis: a ten year follow up study. Br Med J 295:164–168.

NIOSH [1995]. Criteria for a recommended standard: occupational exposure to respirable coal mine dust. Cincinnati, OH: U.S. Department of Health and Human Services, Centers for Disease Control and Prevention, National Institute for Occupational Safety and Health, DHHS (NIOSH) Publication No. 95–106.

NIOSH [2002]. NIOSH hazard review: Health effects of occupational exposure to respirable crystalline silica. Cincinnati, OH: U.S. Department of Health and Human Services, Centers for Disease Control and Prevention, National Institute for Occupational Safety and Health, DHHS (NIOSH) Publication No. 2002–129.

NIOSH [2008]. Work-related lung disease surveillance report, 2007. Morgantown, WV: U.S. Department of Health and Human Services, Centers for Disease Control and Prevention, National Institute for Occupational Safety and Health, DHHS (NIOSH) Publication No. 2008-143a.

Parker JE, Wagner GR [1998]. Silicosis. In: Stellman JM, ed. Encyclopaedia of occupational health and safety. 4th ed. Geneva, Switzerland: International Labour Office, pp. 10.43–10.46.

Peters JM [1986]. Silicosis. In: Merchant JA, Boehlecke BA, Taylor G, Pickett-Harner M, eds. Occupational respiratory diseases. Cincinnati, OH: U.S. Department of Health and Human Services, Centers for Disease Control, National Institute for Occupational Safety and Health, DHHS (NIOSH) Publication No. 86–102, pp. 219–237.

USBM [1992]. Crystalline silica primer. Washington, DC: U.S. Department of the Interior, Bureau of Mines, Branch of Industrial Minerals, Special Publication (SP) 05–92. NTIS No. PB97-120976.

Virta RL [1993]. Crystalline silica: what it is—and isn't. Minerals Today Oct:12–16. Washington, DC: U.S. Department of the Interior, Bureau of Mines.

CHAPTER 2.—SAMPLING TO QUANTIFY RESPIRABLE DUST GENERATION

By Jay F. Colinet[1]

The respirable fraction of the airborne dust is the dust that reaches the lungs and leads to the development of CWP or silicosis. Respirable dust cannot be seen with the eye. Conversely, if a dust cloud is visible, it is likely that a portion of the airborne dust will be in the respirable size range. To quantify the amount of harmful respirable dust in the mine air, sampling instrumentation must be used.

New cases of lung disease in miners have been occurring at increased rates since 2000. As a result, accurate respirable dust sampling is important to quantify worker exposures and identify dust sources. Sampling results can then be used to implement control technologies in the most problematic areas.

RESPIRABLE DUST SAMPLERS FOR USE IN COAL MINING

The most common type of sampler used in the mining industry is the gravimetric sampler (Figure 2-1). This device is designated for use in compliance dust sampling by the Federal Coal Mine Health and Safety Act of 1969. It consists of a constant-flow sampling pump, a size-selective cyclone, and a filter cartridge. For coal mining operations, the sampling pump should be calibrated to operate at 2 lpm. In metal/nonmetal mining operations, the pump should be operated at 1.7 lpm. The 10-mm Dorr-Oliver cyclone separates the oversize dust from the respirable fraction (usually considered to have an aerodynamic diameter of 10 µm or less). The oversize dust is deposited into the grit pot at the bottom of the cyclone, while the respirable fraction is deposited onto a 37-mm-diam polyvinyl chloride (PVC) filter. The filter collects the respirable dust and should be weighed by a qualified lab to determine the mass of dust that has been collected during sampling. The mass of dust and the volume of air sampled are used to calculate the concentration of respirable dust in milligrams per cubic meter. Care must be taken after a sample is collected to ensure that the cyclone assembly stays in an upright position. Otherwise, the oversize dust particles in the grit pot can be deposited onto the filter and invalidate the sample.

---

[1]Supervisory mining engineer, Office of Mine Safety and Health Research, National Institute for Occupational Safety and Health, Pittsburgh, PA.

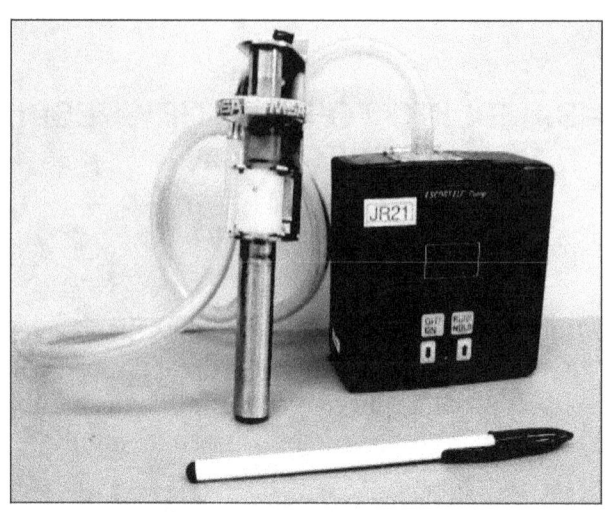

Figure 2-1.—Gravimetric sampling pump, cyclone, and filter cassette.

To determine the silica content of a gravimetric sample, the filter must be sent to an accredited laboratory for analysis. For samples collected in coal mines, the MSHA P7 infrared analytical technique [Parobeck and Tomb 2000] is used to determine silica content. For samples collected in metal/nonmetal mines, x-ray diffraction using NIOSH Method 7500 [Schlecht and O'Connor 2003] is used.

Because of the great number of variables encountered in mining operations that can impact airborne dust levels, it is highly desirable to place multiple gravimetric samplers at a single location and calculate an average dust concentration. The use of multiple samplers increases the confidence that the measured dust levels are representative of the true dust concentration.

In addition to gravimetric samplers, a real-time dust sampler has been approved by MSHA for use in underground mines, but not for compliance sampling purposes. The personal DataRAM (pDR) has dust-laden air pass through a sensing chamber and passes a light beam through the dust. A sensor measures the amount of light scatter caused by the dust and relates this scatter to a relative dust concentration. This concentration is correlated to the time when the sample was measured and stores this information in the internal data logger. The sample data can then be downloaded to a computer for analysis. Figure 2-2 illustrates a typical graph obtained with the pDR, as well as a photo of the pDR. Mobile sampling was used to collect the data (this sampling technique will be discussed in the next section). The time-related dust data can be analyzed for specific time intervals (e.g., head-to-tail passes on longwalls), with average dust concentrations calculated for these intervals.

Unfortunately, the accuracy of the light-scattering monitor can be compromised by dust clouds with different size distributions, different dust compositions, and/or water mist in the air. Consequently, when NIOSH uses pDR samplers, a field calibration is completed. Gravimetric samplers are placed adjacent to the pDR, and individual pDR dust measurements are adjusted based on the ratio between the average gravimetric concentration and the average pDR concentration [Thermo Scientific 2008]. For example, if the gravimetric concentration was 1.3 mg/m^3 over a 6-hr measurement period and the pDR average concentration was 1.0 mg/m^3 for the same 6 hr, then all individual pDR measurements would be multiplied by 1.3.

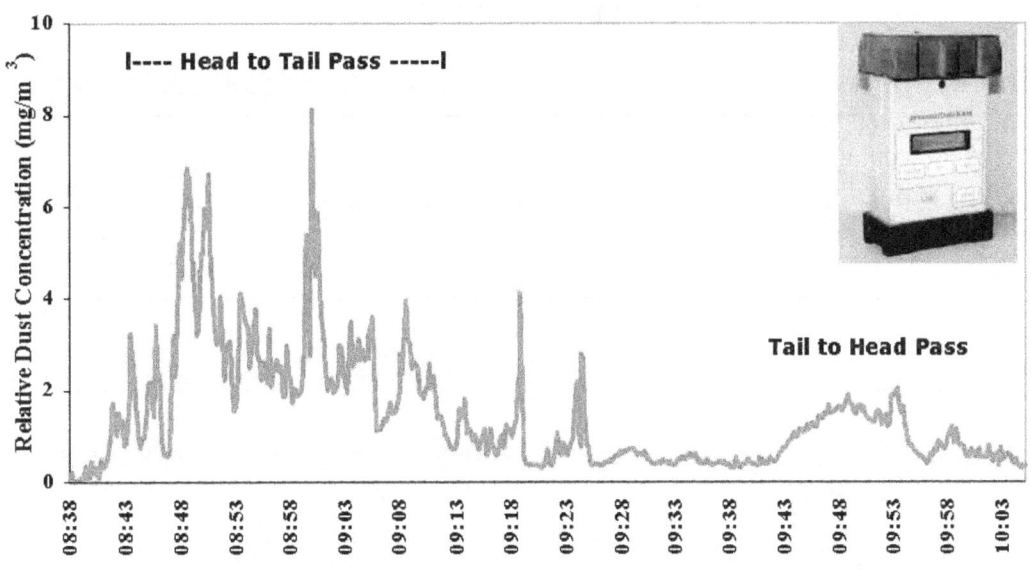

Figure 2-2.—Example of dust measurements obtained with the pDR.

The personal dust monitor (PDM) is another real-time sampler that has been developed and tested by NIOSH, approved for use in underground coal mines by MSHA, and reached commercial production [Volkwein et al. 2006]. The PDM uses the tapered-element oscillating microbalance (TEOM) to obtain a real-time, gravimetric-based measure of respirable dust concentrations. The TEOM is a hollow tube that vibrates at a known frequency with a filter mounted on the end. As respirable dust is deposited onto the filter, the TEOM frequency changes, which can be related to a dust concentration. The PDM provides the wearer with a readout that displays the cumulative dust concentration to that point in the shift and the percent of the permissible exposure limit that has been reached. This information can be used by the wearer to monitor dust exposure during the shift to prevent overexposure. The sampler is incorporated into standard cap lamp housing and has the sampling inlet located at the cap lamp (Figure 2-3).

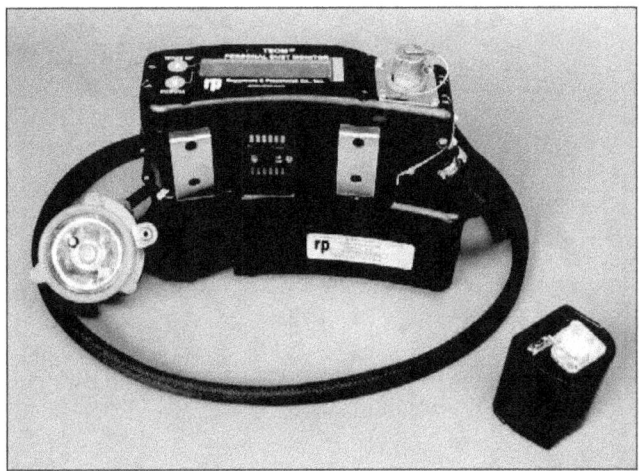

Figure 2-3.—PDM with TEOM removed *(shown on right)*.

SAMPLING STRATEGIES

To effectively control the respirable coal and silica dust exposure of mine workers, it is necessary to identify the sources of dust generation and quantify the amount of dust liberated by these sources. Once the dust sources are identified and quantified, dust control technologies that offer the greatest protection to the mine workers can then be applied.

To quantify the amount of dust liberated by a source, dust sampling must be conducted in a manner that isolates the identified dust-generating source. This is achieved by placing dust samplers upwind and downwind of the source in question. The difference between these measurements is used to calculate the quantity of dust liberated by this source.

For example, in an underground coal mine, samplers can be placed in the immediate intake and return of the continuous miner to determine the amount of dust liberated by the miner while cutting and loading in the face. In this case, samplers are positioned upwind and downwind of the miner to sample the airborne dust levels throughout the cut. Figure 2-4 shows these sampling locations. If gravimetric samplers are used for this evaluation, it will be necessary to ensure that sufficient mass is collected during sampling. As a result, it may be necessary to sample during multiple continuous miner cuts. In this case, the sampling pumps should be started when the continuous miner has been positioned in the face and begins cutting the coal. After the first cut has been completed, the sampling pumps are turned off during the time the miner is repositioning into the next face. While off, the sampling pumps should be repositioned into the second cut in the same relative locations as for the first cut sampled. When the miner is ready to resume mining, the sampling pumps can be restarted.

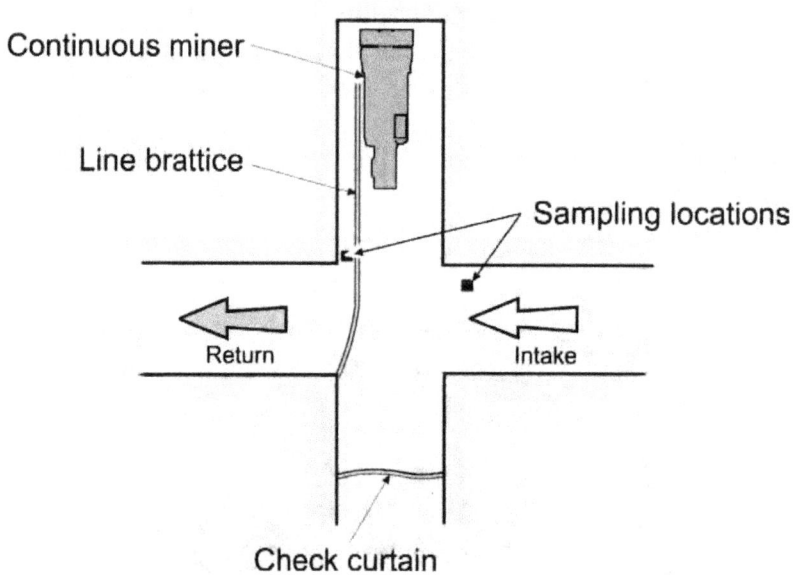

Figure 2-4.—Sampling locations used to isolate dust generated by a continuous miner.

For a more mobile piece of equipment, such as a longwall shearer, a mobile sampling strategy must be used to isolate the dust generated by the equipment. Two sampling personnel would be required to travel with the shearer as it mines across the longwall face. One person would be located upwind of the shearer, while the second would be located downwind. These sampling personnel would maintain their respective distances from the shearer as it mines across the face. Figure 2-5 illustrates this mobile sampling strategy.

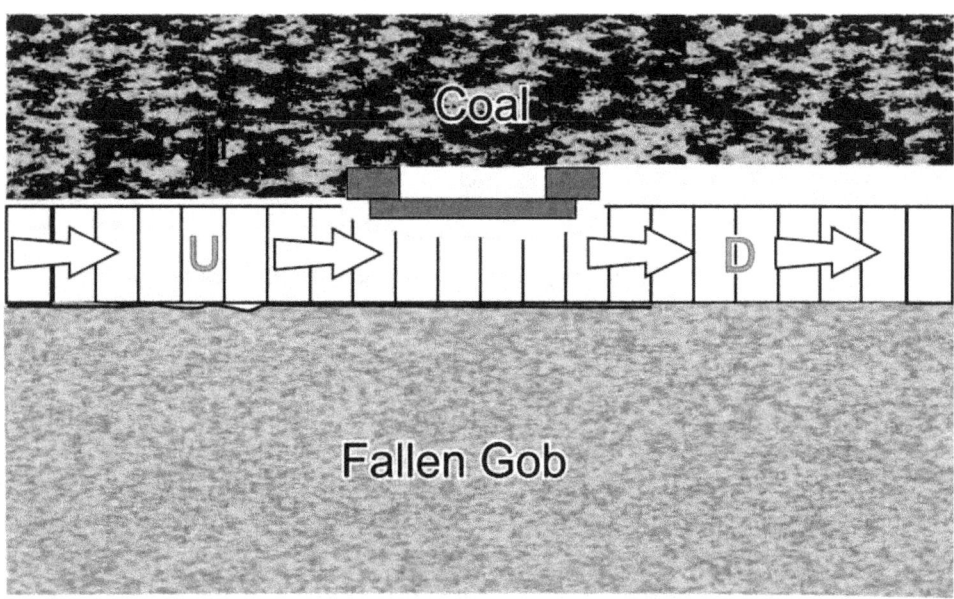

U - Upwind location   D - Downwind location

Figure 2-5.—Mobile sampling used to quantify shearer dust.

Both of these sampling examples represent underground coal mines where a well-defined ventilation pattern is typically present. However, this is not always the case. For example, to quantify the amount of respirable dust generated by a drill at a surface mine, it would be necessary to place an array of samplers around the drill to account for dust liberated during changing wind directions. The dust concentrations from these samplers would be averaged to quantify dust liberation around the drill. It would also be necessary to place a background dust sampler far enough away from the drill, so that it is not impacted by drill dust, to monitor ambient dust levels. The dust levels from the ambient sample would be subtracted from the drill samples that have been averaged to determine dust liberated by the drill. Figure 2-6 shows sampling locations around a surface drill.

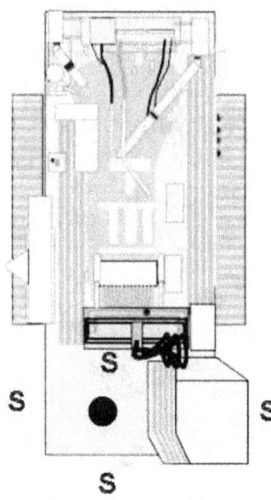

A – Ambient sampling location   S – Drill sampling locations

Figure 2-6.—Sampling locations around a surface drill.

After identifying and quantifying the most significant dust sources, appropriate dust controls should be selected and implemented. To determine the impact of the added controls, sampling would once again be conducted. Typically, an A-B comparison would be needed to quantify the impact of added control technologies. The A-portion of the sampling would be conducted with the original operating conditions to establish baseline dust levels. The control technology of interest would then be installed and the B-portion of the testing completed. To maximize the validity of the test results, both portions of the testing should be completed under similar operating conditions. The dust levels measured under each test condition would be compared to quantify the effectiveness of the installed control.

REFERENCES

Parobeck PS, Tomb TF [2000]. MSHA's programs to quantify the crystalline silica content of respirable dust samples. SME preprint 00-159. Littleton, CO: Society for Mining, Metallurgy, and Exploration, Inc.

Schlecht PC, O'Connor PF, eds. [2003]. NIOSH manual of analytical methods (NMAM®), 4th ed., 3rd supplement. Cincinnati, OH: U.S. Department of Health and Human Services, Centers for Disease Control and Prevention, National Institute for Occupational Safety and Health, DHHS (NIOSH) Publication No. 2003–154.

Thermo Scientific [2008]. Model pDR-1000AN/1200 instruction manual. Franklin, MA: Thermo Scientific, pp. 35–36.

Volkwein JC, Vinson RP, Page SJ, McWilliams LJ, Joy GJ, Mischler SE, Tuchman DP [2006]. Laboratory and field performance of a continuously measuring personal respirable dust monitor. Pittsburgh, PA: U.S. Department of Health and Human Services, Centers for Disease Control and Prevention, National Institute for Occupational Safety and Health, DHHS (NIOSH) Publication No. 2006–145, RI 9669.

CHAPTER 3.—CONTROLLING RESPIRABLE DUST ON LONGWALL MINING OPERATIONS

By James P. Rider[1] and Jay F. Colinet[2]

Medical studies have shown that prolonged exposure to excessive levels of respirable coal dust can lead to coal workers' pneumoconiosis (CWP), progressive massive fibrosis, and chronic obstructive pulmonary disease. These lung diseases are irreversible and can be debilitating, progressive, and fatal. CWP contributed to the deaths of 10,406 U.S. miners during 1995–2004 [NIOSH 2008]. Pneumoconiosis continues to be a very serious health threat to underground coal mine workers.

Historically, longwall operations have had difficulty in maintaining consistent compliance with the federal dust standard of 2.0 mg/m^3. During 2004–2008, mine operators and MSHA inspectors collected 6,600 and 1,321 valid compliance samples, respectively, from longwall designated occupations or high-risk occupations. These dust samples showed that 719 (11%) of the mine operator samples and 144 (11%) of the MSHA samples exceeded 2.1 mg/m^3 [Niewiadomski 2009]. In addition, MSHA inspector sampling results from 2004–2008 show that longwall face workers were exposed to elevated levels of respirable silica dust. For MSHA occupation codes 044 (tail-side shearer operator) and 041 (jack setter) that were on reduced dust standards due to silica levels above 5%, 31% and 21% of the samples, respectively, exceeded the reduced standard [MSHA 2009]. The continued occurrence of CWP in underground coal mine workers and the magnitude of respirable dust overexposures in longwall mining occupations point to the need for improved dust control technology on longwalls.

Longwall mining equipment and operational practices have improved dramatically since the early 1980s. In 2007, longwall mines accounted for 50% of U.S. underground coal production. Overall production from U.S. longwall mines peaked in 2004 and decreased by about 10% in 2007 with over 176 million tons mined [EIA 2009]. These production rates continue to challenge dust control efforts of the industry.

Longwall workers can be exposed to harmful respirable dust from multiple dust generation sources, including the intake entry, belt entry, stageloader/crusher, shearer, and shield advance. This chapter discusses dust control technologies that are available to reduce dust liberated from each of these sources. Alternate controls that have the potential to provide additional dust reductions but currently not in use will also be discussed.

CONTROLLING RESPIRABLE DUST ON INTAKE ROADWAYS

Respirable dust concentrations outby the face area in intake roadways may have a significant effect on dust exposures of longwall face workers if not properly addressed. Recent longwall dust surveys revealed that respirable dust levels in the last open crosscut can be as high as 0.42 mg/m^3 [Rider and Colinet 2007]. Also, as longwall production has increased, mine operators are bringing larger quantities of air to the face to control methane and dust liberation.

---

[1]Operations research analyst.
[2]Supervisory mining engineer.
Office of Mine Safety and Health Research, National Institute for Occupational Safety and Health, Pittsburgh, PA.

Average air quantities on the longwall faces are higher than ever and increased about 65% compared to levels from a longwall study in the mid-1990s [Colinet et al. 1997].

Higher air velocities in the intake entries may result in increased dust entrainment if proper controls are not applied. Increasing air velocities have been shown to have the potential to entrain greater quantities of dust if sufficient moisture is not present. NIOSH studies [Listak et al. 2001; Chekan et al. 2001, 2004] quantified increased entrainment when the dust was dry (1% moisture or less) and falling into the ventilating airstream, similar to dust dropping into the air during shield advance. Consequently, activities that disturb dry dust on the intake roadways may contribute to dust reaching the longwall face.

The following practices can help control respirable dust levels along intake roadways:

- **Limit support activities during production shifts.** Vehicle movement, removal of stoppings, and delivering/unloading supplies during production shifts can elevate intake dust levels. These activities combined with increased air velocities can cause dust to be entrained into the face ventilating airstream, especially if they occur close to the last open crosscut.

- **Apply water or hydroscopic compounds to control road haulage dust.** Water application to the mine floor is crucial to control respirable dust in the intake roadway. Operators must be diligent in monitoring moisture content of the dust along intake roadways, especially with the increased amount of air traveling toward the face and during winter months. This air amplifies the potential for the roadways to dry out more quickly. The moisture content of the haulage floor should be approximately 10% [Organiscak and Reed 2004; Kost et al. 1981]. Hydroscopic compounds such as calcium, magnesium chloride, hydrated lime, and sodium silicates increase roadway surface moisture by extracting moisture from the air. Applications of these materials will help maintain the moisture content of the road surface [Organiscak et al. 2003].

- **Use surfactants.** Surfactants such as soaps and detergents dissolve in water and can be beneficial in maintaining the proper moisture content of the intake roadways. Surfactants decrease the surface tension of water, which allows the available moisture to wet more particles per unit volume [Organiscak et al. 2003].

CONTROLLING RESPIRABLE DUST FROM THE BELT ENTRY

Using the belt entry to complement the intake entry will allow the delivery of more air to the face, providing the potential for better dust and methane dilution. Recent longwall surveys showed that about 40% of the operations were using belt entry air [Rider and Colinet 2007]. Compliance data analyzed by MSHA [1989] showed that mines using belt air to ventilate work areas did not have significantly different respirable dust levels at the designated occupations when compared to the mines not using belt air. Also, studies conducted by the U.S. Bureau of Mines [Potts and Jankowski 1992; Jankowski and Colinet 2000] indicated that any potential addition to dust levels at the longwall face from the belt entry seems to be mitigated as a result of the increased dilution that can be obtained with additional air brought up the belt entry.

However, the potential for dust from the belt entry to contaminate the face area has increased in recent years because the quantity of coal being transported by the belt continues to increase. The following practices can help control respirable dust levels in the belt entry:

- **Belt maintenance.** Properly maintaining the belts is one of many vital operating practices necessary to keep respirable dust levels low along the belt entry. Missing rollers, belt slippage, and worn belts can cause belt misalignment and create spillage [Organiscak et al. 1986]. Given the increases in the quantity of coal being transported outby the face, operators must be diligent in their efforts to properly maintain the existing belt entry dust suppression controls to keep fugitive dust from being entrained and carried by the ventilation airstream to the face area.

- **Wetting the coal product during transport.** If the coal is wetted adequately at the face, less dust will be created during transport at the transfer points. However, with the substantial increase in airflow in the belt entry, the moisture may evaporate and rewetting of the coal may be necessary at multiple intervals along the belt. Flat-fan sprays and full-cone nozzles are typically used for coal wetting along the belt. Water application usually ranges from 1 to 4 gpm at operating pressures of about 50 psi [Kost et al. 1981].

- **Belt cleaning by scraping and washing.** Scraping and washing of the belt play an important role in reducing the amount of dust generated by the conveyor belt [Kissell and Stachulak 2003; Organiscak et al. 1986; Shirey et al. 1985]. Material that adheres to the belt is subject to crushing at the head and tail roller. Often this material dries out and becomes airborne as it passes over the return idlers. The top and bottom of the return belt should be cleaned with spring-loaded or counterweighted scrapers. A low-quantity water spray may be necessary to moisten the belt slightly and complement the belt scrapers. Previous studies [Stahura 1987; Baig et al. 1994] have shown that water sprays in conjunction with belt scrapers significantly reduced airborne respirable dust levels.

- **Use of a rotary brush that cleans the conveying side of the belt.** A motor-driven rotary brush [Organiscak et al. 1986] that cleans the conveying side by rotating in the opposite direction of the conveyor belting (Figure 3-1) will help reduce dust levels along the belt. This brush should be located near the dump point so that the material sticking to the belt is still wet and agglomerated as it is brushed off. As the material gets carried back on the belt return, it can dry and become airborne when dislodged from the belt.

- **Wetting of dry belts.** Studies have shown that wetting the bottom (nonconveying-side) belt can significantly reduce dust from a dry belt as it returns from the dump point [Kissell and Stachulak 2003; Organiscak et al. 1986; Shirey et al. 1985]. A full-cone water spray is directed onto the nonconveying side of the belt (which is the top side as the belt returns), followed by a piece of material such as a foam-backed piece of carpet positioned across the width of the belt to wipe the belt and remove the dust fines (Figure 3-2).

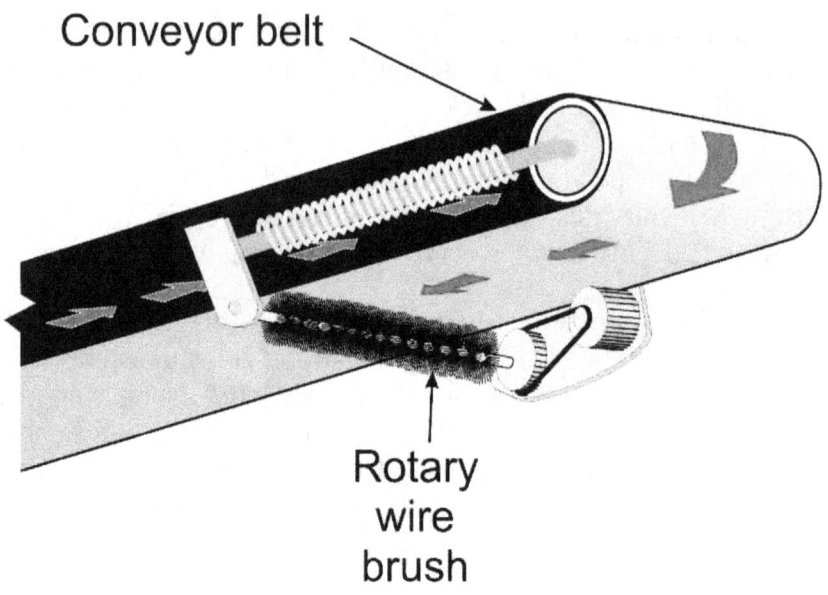

Figure 3-1.—Rotary brush cleans the conveying side of the belt.

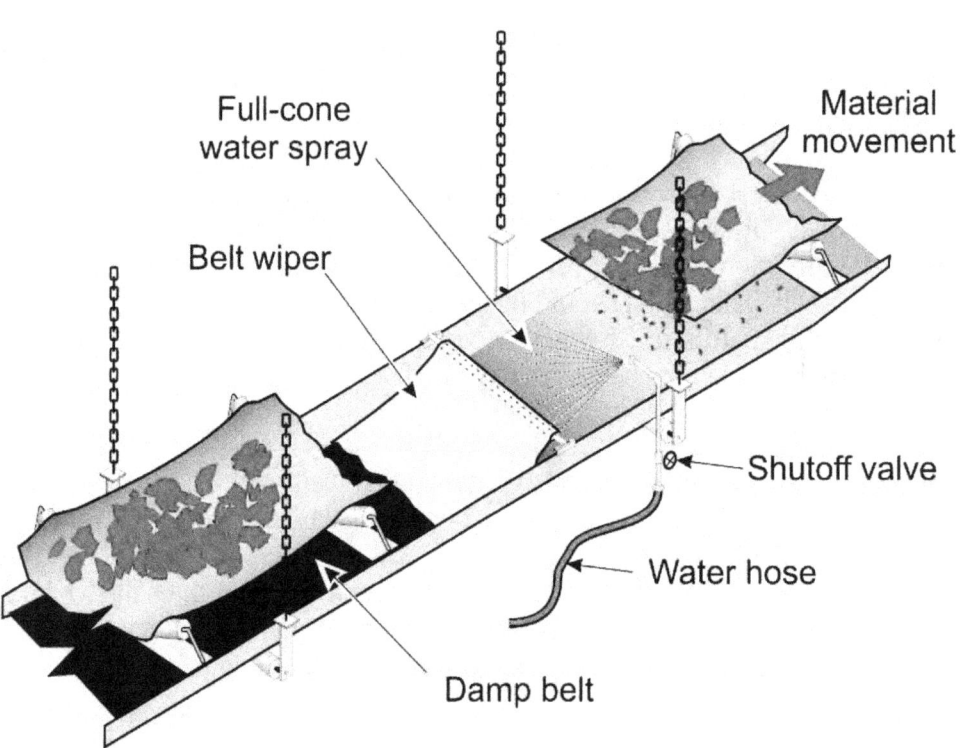

Figure 3-2.—Water sprays and belt wiper used to reduce dust from the nonconveying side of the belt as it returns.

CONTROLLING RESPIRABLE DUST IN THE HEADGATE ENTRY, INCLUDING THE STAGELOADER/CRUSHER

Respirable dust generated by outby sources can enter the ventilating airstream and remain airborne across the entire longwall face, which can impact the dust exposure of all personnel on the face. The stageloader/crusher is the most significant dust-generating source in the headgate area. The breaking of coal and rock in the crusher generates large quantities of dust, which can mix with the ventilating airstream.

The following practices can help control respirable dust levels in the stageloader/crusher area:

- **Fully enclosing the stageloader/crusher.** Recent NIOSH longwall surveys [Rider and Colinet 2007] found that all stageloader/crushers were fully enclosed. However, there was not a universally applied technique for enclosing the stageloader/crusher. The common practice is to apply a combination of steel plates, strips of conveyor belting, brattice, and/or foam to seal the crusher and stageloader units along their entire length. In addition, conveyor belting covering the entrance of the crusher has been effective in keeping dust from boiling out of the enclosure and into the ventilating airstream. Strips of belting were hung from the top of the crusher inlet, effectively enclosing this area. With the quantity of coal being transported through the stageloader/crusher, it is imperative that all seals and skirts be carefully maintained to confine dust generated within the enclosure.

- **Wetting the coal in the crusher and stageloader area.** Crushers should have a built-in spray manifold located above the crusher hammers. Traditional water flow to this manifold is 8–10 gpm. In addition, a spray manifold consisting of three or four full-cone sprays is typically mounted at the entrance to the crusher's enclosure [Jankowski and Colinet 2000; Organiscak et al. 1986; Shirey et al. 1985]. The spray bar should span the width of the conveyor to ensure uniform spray coverage. The objective of these sprays is to wet the coal product and prevent respirable dust from becoming airborne. Previous studies [USBM 1985; Kelly and Ruggieri 1990] have shown that low water pressure and high-volume sprays are the most effective at containing dust within the enclosure. High-pressure sprays should be avoided since they may force dust out of the enclosure and into the ventilating airstream. Because water quantity is more critical than water pressure, the use of larger-orifice, full-cone sprays operating at water pressures below 60 psi is recommended.

  Often, a spray bar is located at the discharge of the crusher. A spray bar located above the belt immediately at the stageloader-to-belt transfer point can also be used to reduce dust levels at this transfer point [Organiscak et al. 1986; Shirey et al. 1985; USBM 1985]. Figure 3-3 shows the various locations of sprays that are recommended.

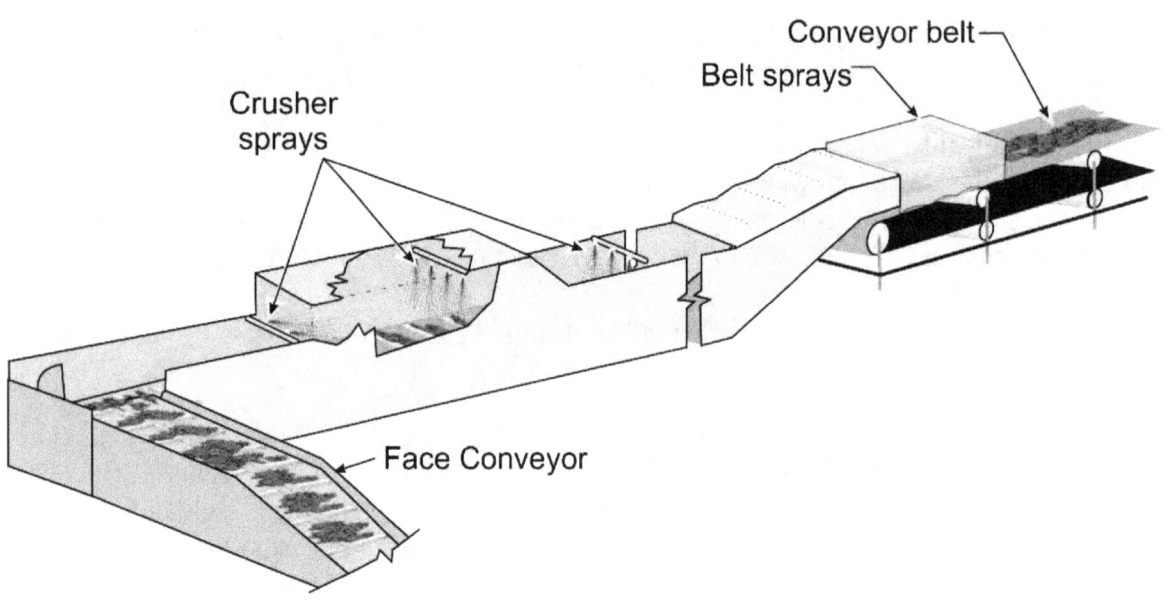

Figure 3-3.—Enclosed stageloader/crusher and location of water sprays.

- **Using scrubber technology in the stageloader/crusher area.** In an effort to keep fugitive dust from escaping the stageloader/crusher area, fan-powered scrubbers located close to the crusher discharge and/or stageloader-to-belt transfer area can be used. If scrubbers are used, their inlets are commonly ducted from the crusher discharge area and the stageloader-to-belt transfer. Flow rates through the scrubber typically range from 6,500 to 8,500 cfm. In addition to capturing airborne dust, the scrubber also creates a negative pressure within the enclosed stageloader/crusher to minimize dust from leaking out if any gaps are present.

- **Using a high-pressure water-powered scrubber.** A compact, high-pressure, water-powered scrubber is an alternative to fan-powered scrubbers [Kelly and Ruggieri 1990]. A water spray installed at the center of a tube and operated at pressures of at least 1,000 psi will induce airflow through the tube as well as capture most of the dust in the airflow [Jayaraman et al. 1981]. Since this scrubber is water-powered, it is intrinsically safe in relation to methane, and maintenance requirements are minimal because the scrubber has no moving parts. Successful underground tests were conducted where contaminated air was drawn through a series of five tubes with sprays attached to each tube. The dirty air was scrubbed through the tubes and demisted through a wave-blade demister. Figure 3-4 shows the scrubber mounted on top of the crusher. Cleaned air was discharged toward the face. Field tests showed that the scrubber reduced dust concentrations by more than 50% when operated at 1,200 psi and 10 gpm.

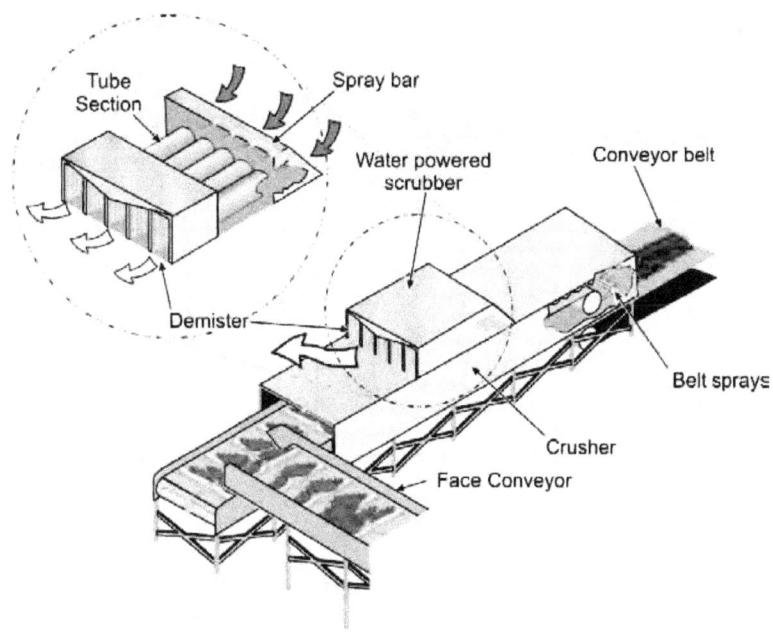

Figure 3-4.—High-pressure water scrubber installed on top of crusher.

In addition to the stageloader/crusher controls, the following practices at the headgate can help reduce the dust exposure of longwall face workers:

- **Installation and maintenance of a gob curtain.** Adequate ventilation of the longwall panel involves supplying the required volume of air to the headgate and maintaining that airflow along the face. Often, loss of air into the gob in the headgate area prevents the maximum utilization of the air intended to ventilate the longwall face. As a result of roof bolting in the belt entry, the roof behind the shield supports may not collapse as quickly as it does along the rest of the face. This can result in a larger opening behind the first few shields and allow a substantial portion of the ventilating air delivered to the headgate to leak into the gob. The open area between the first shield and the rib also facilitates leakage into the gob. Furthermore, fresh air traveling into the gob may become contaminated with dust and may reenter the face area, compounding the dust problem. A gob curtain (Figure 3-5) installed between the first support and the rib in the headgate entry can force the ventilating air to make a 90° turn down the longwall face rather than leak into the gob. A number of longwall operations have installed brattice curtain behind the hydraulic support legs along the first 5–10 shields in an effort to further reduce leakage into the gob and increase airflow down the face.

  In various studies, the average face air velocity with the curtain installed was about 35% greater than without the curtain. The biggest improvement due to the curtain was seen at the first 25–30 supports, where increased air volume lowered dust concentrations through dilution [Kissell et al. 2003; Jankowski and Colinet 2000; Shirey et al. 1985]. All of the recent NIOSH longwall surveys found that gob curtains were being used in the headgate entry. Unfortunately, most of the curtains were not properly maintained, resulting in large voids with air escaping into the gob.

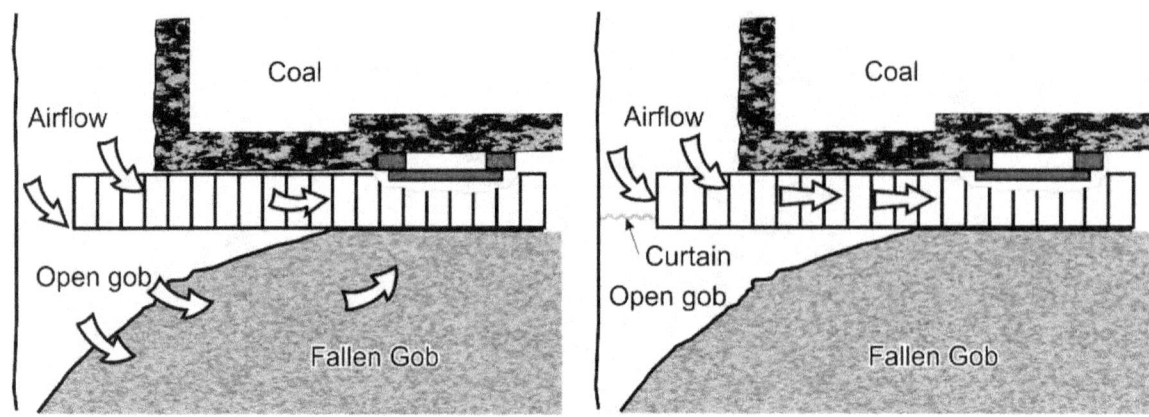

Figure 3-5.—Gob curtain increases airflow down the face.

- **Positioning shearer operators outby as the headgate drum cuts into the headgate entry.** One source of elevated dust concentrations for shearer operators is when the headgate drum cuts into the headgate entry. The drum is exposed to the primary airstream, resulting in air passing over and around the cutting drum. The air picks up large quantities of respirable dust, potentially exposing the shearer operators. Although the cutout time is relatively short, the dust levels inby the headgate drum and typically where shearer operators are located can be high and have been observed in the range of 20–30 mg/m^3 [Jankowski and Colinet 2000; Shirey et al. 1985]. In recent NIOSH dust surveys [Rider and Colinet 2007], a concerted effort was made by both the headgate and tailgate shearer operators at many longwalls to move outby the shearer headgate drum prior to the drum cutting out into the entry. Typically, they positioned themselves behind the face conveyor drive motors near shields 1 and 2, which is upwind of the headgate drum and also offers protection from flying coal. Locating the shearer operators in this area can reduce their exposure to potentially high dust concentrations as the headgate drum cuts into the headgate entry.

- **Installation of a wing or cutout curtain between the panel side rib and the stageloader.** In addition to locating the shearer operators outby the contaminated airstream when cutting into the headgate entry, installing a wing curtain [Jankowski and Colinet 2000; Shirey et al. 1985] can reduce dust entrainment (Figure 3-6). The curtain is suspended from the roof between the panel-side rib and the stageloader. Previous research [Jankowski et al. 1986] has shown that a wing or cutout curtain is effective in reducing downstream exposure levels. The curtain shields the headgate drum as it cuts out into the headgate entry, directing the airflow around the drum. It should be located 6 ft from the corner of the face to provide maximum shielding without interfering with the drum.

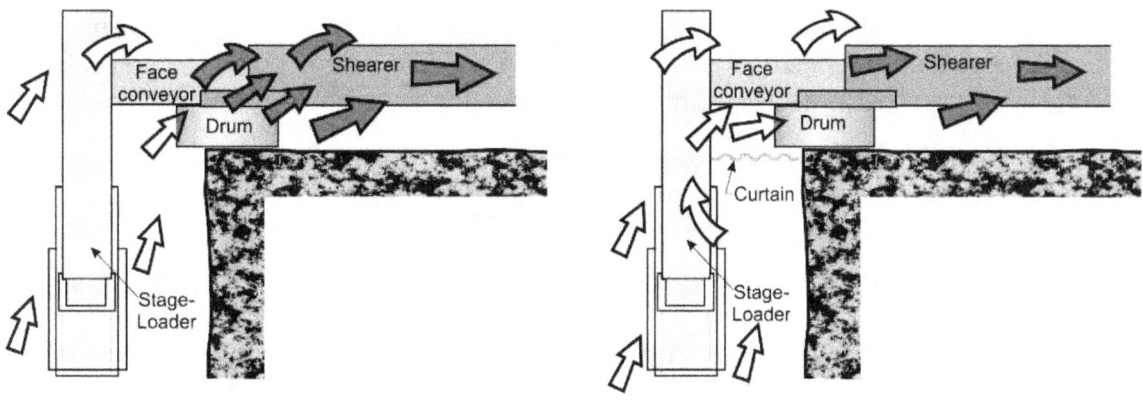

Figure 3-6.—Ventilation patterns around shearer without *(left)* and with *(right)* a cutout curtain.

CONTROLLING SHEARER DUST

On most longwall faces, the cutting action of the shearer is the primary dust source and the largest contributor to the respirable dust exposure of face personnel. Therefore, shearer generated dust should be the major focus of any control effort, especially if a bidirectional cutting sequence is used. Previous longwall studies showed the shearer was the largest source of dust on longwall panels compared with intake, stageloader, and shield support dust sources [Colinet et al. 1997]. It accounted for over 50% of all dust generated during mining. Following is a discussion of a number of technologies for controlling shearer-generated dust.

- **Face ventilation.** As with all mining methods, ventilation is the primary means to dilute liberated methane to safe levels. It is also the principal method of controlling respirable dust on the longwall face. Providing adequate amounts of air to dilute and carry airborne dust down the face and prevent it from migrating into the walkway has been and continues to be a goal for longwall operators. Previous studies [Mundell et al. 1979] have reported that face air velocities of 400–450 fpm seem to be the minimum appropriate to control respirable dust. A German study [Breuer 1972] reported that the optimum velocity range may be increased to 700–900 fpm when the moisture content of the dust particles is 5%–8%. An MSHA study [Tomb et al. 1992] reported that as face air quantities increased, even beyond 1,200 fpm, respirable dust levels along the face decreased. As air velocities increase, it is important to ensure that sufficient wetting of the coal is provided to minimize the potential of increased entrainment with the higher air velocities. The higher velocities provide greater air quantities for better dilution of intake dust, as well as dust generated during shield support movement. Higher velocities over the shearer help confine the dust to the face area and lower the potential for contaminating the walkway. Also, the higher velocities will improve the diffusion of dust from stagnant areas in the headgate and along the support line. In recent NIOSH surveys [Rider and Colinet 2007], the average velocity was 665 fpm, and two longwalls had velocities over 800 fpm. The average quantity of air along the face was 67,000 cfm, an increase of about 65% compared to the mid-1990s longwall study [Colinet et al. 1997]. Also, dust levels measured upwind of the shearer and at shearer midpoint were lower in the recent surveys [Rider and Colinet 2007] compared with those of earlier studies. This suggests that the increase in air velocity along with the use of the shearer directional spray systems confines the shearer dust close to the face and prevents it from migrating into the walkway.

- **Drum-mounted water sprays.** Drum-mounted water sprays apply water for dust suppression directly at the point of coal fracture and add moisture to the product to minimize dust liberation during coal transport. Although very effective at minimizing dust generation at the point of coal fracture, shearer drum water sprays can actually increase airborne respirable dust levels if operated at water pressures that are too high. Instead of suppressing dust generation, these sprays can force the dust out away from the cutting drum, allowing it to mix with the primary airflow, where it is then carried throughout the entire cross-sectional area of the longwall face [Jankowski and Colinet 2000]. Previous studies [Shirey et al. 1985] have shown that shearer drum water sprays are very effective at minimizing dust generated, but increasing shearer drum water spray pressure above 100 psi can increase the shearer operator's dust exposure by as much as 25%. For most operations, the optimum operating drum spray pressure seems to be 80–100 psi. Full-cone sprays are the most effective type of spray pattern to use in shearer drums. These sprays increase wetting without inducing substantial air movement around the drum. Reducing nozzle pressures while increasing water quantity can be accomplished by installing spray nozzles with larger orifices that provide greater flow at reduced operating pressures.

- **Cutting drum bit maintenance.** Previous research has shown that bits with large carbide inserts and a smooth transition between the steel shank and the carbide reduce dust levels [Organiscak et al. 1996]. The prompt replacement of damaged, worn, or missing bits cannot be overemphasized. A dull bit rubs against the coal, which results in an ineffective use of the available cutting force and the inability to penetrate the coal at designed rates. This results in shallow cutting, which greatly increases dust generation. Not only do dull bits result in higher cutting forces and more dust, but there is also an increased likelihood for mechanical damage of bit holders and gear boxes and for frictional ignition of methane [Shirey et al. 1985].

- **Directional water spray systems.** Water sprays can be very efficient air movers and, if applied properly, can be used to augment the primary airflow and reduce the amount of shearer-generated dust that migrates into the walkway near the shearer. Water sprays mounted on the shearer body act very much like small fans, moving air and entraining dust in the direction of their orientation [Jankowski and Colinet 2000]. Poorly designed shearer-mounted spray systems with nozzles directed upwind at the cutting drums actually force dust away from the face, where it mixes with clean intake air and is carried out into the walkway over the shearer operators. A directional spray system called the shearer-clearer [Jayaraman et al. 1985] takes advantage of the air-moving capabilities of water sprays and confines the dust-laden air against the face. It consists of several shearer-mounted sprays oriented downwind to augment the primary ventilation airflow. Also, it includes one or more passive barriers that split the airflow around the shearer into clean and contaminated air (Figure 3-7). The air split is initiated by a splitter arm that extends from the walkway side of the shearer body parallel to the headgate ranging arm. Conveyor belting hangs down from the splitter arm to the pan line to provide a physical barrier between the face conveyor and the walkway. In addition, a series of water sprays is mounted on top of the splitter arm to induce airflow and dust movement toward the coal face.

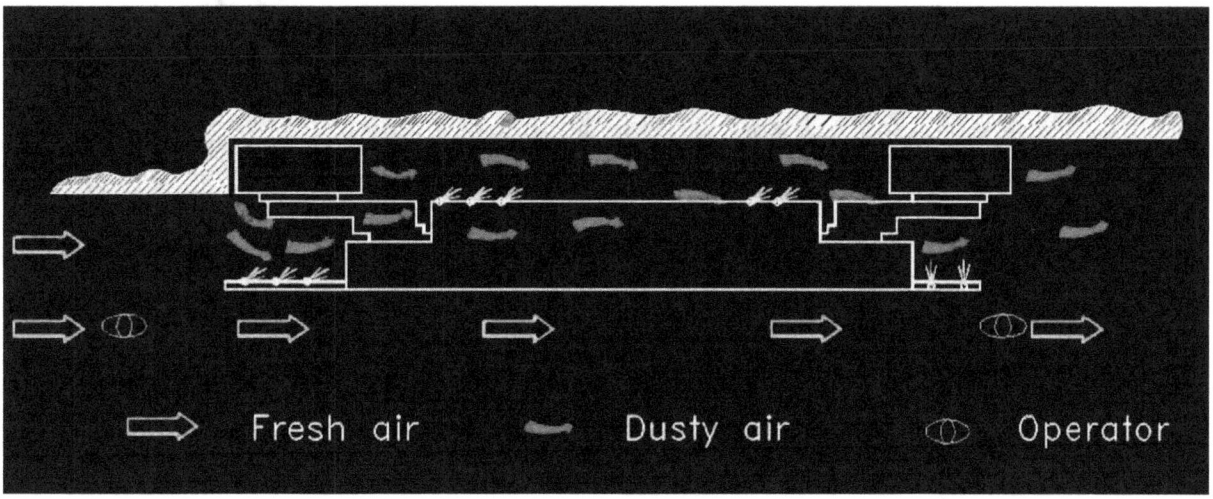

Figure 3-7.—Shearer-clearer directional spray system.

To maximize the effectiveness of the shearer-clearer system, the splitter arm should extend as far beyond the headgate drum as possible, all splitter arm sprays should be oriented with the airflow, a sufficient number of sprays should be used to prevent dust from the headgate drum from migrating to the walkway, and belting should be hung from the splitter arm to help separate face airflow and confine dust. Since the splitter arm should extend beyond the drum if possible, it should be made from sufficiently rigid steel tubing/pipe to withstand coal and rock impacts from the face. Alternately, splitter arms have been observed where springs have been mounted on the arm so that the arm can absorb a blow and bounce back into position. Since directional spray systems are attempting to move air, the operating pressure is critical and pressures of at least 150 psi should be used. Hollow-cone or venturi sprays (Figure 3-8) are effective for these systems. The sprays should be oriented to help move dust along the face without causing turbulence. Thus, it is not desirable to have sprays impacting the ranging arm.

Figure 3-8.—Venturi sprays mounted on headgate splitter arm.

Conveyor belting suspended along the length of the splitter arm, along with the directional sprays, helps split the airflow coming down the face. The belting also provides a physical barrier between the face conveyor and walkway, which helps prevent dust from moving into the walkway. Tears and gaps in the conveyor belting greatly compromise the effectiveness of the splitter arm. Locating sprays on the walkway side of the splitter arm and directing the sprays down the side of the belting (Figure 3-9) may help limit dust migration into the walkway. High-capacity, low-pressure, flat-fan sprays spaced evenly along the length of the splitter arm and directed down the side of the belting can block any fugitive dust from escaping beyond the splitter arm.

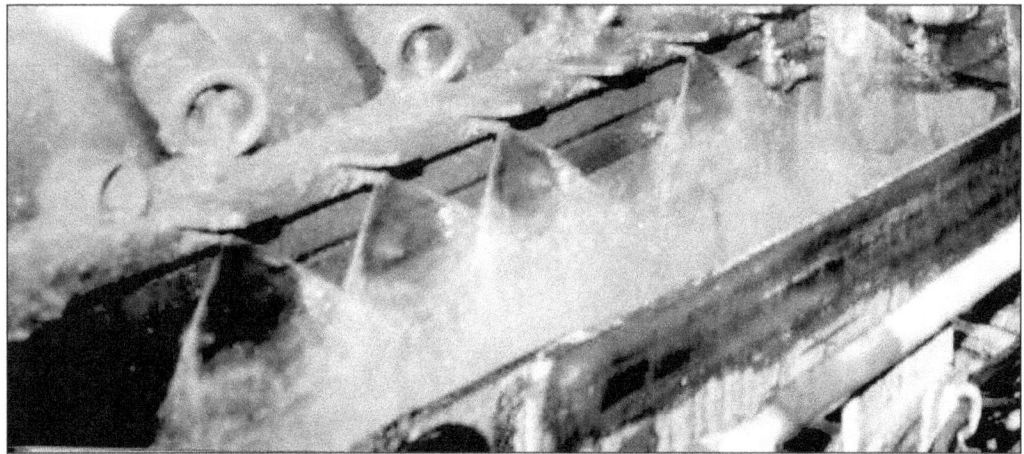

Figure 3-9.—Headgate splitter arm with flat-fan sprays mounted on gob side of belting.

An alternative to the walkway-side fan sprays would be to mount sprays on the underside of the splitter arm. Once again, high-capacity fan sprays could be positioned evenly along the length of the splitter arm and aimed down toward the conveyor. These sprays may have a positive effect on reducing the dust rolling under or through the splitter arm belting and should add more water to the coal product on the pan line, thus reducing conveyor dust. Achieving the desired results by locating sprays on the underside of the splitter bar may be challenging given the amount of turbulence in that area. Spray pressure becomes critical, and low spray pressure may not be effective in reducing the dust migrating under the belt, while too high of a spray pressure may create more turbulence at the bottom of the belting and induce more dust to migrate into the walkway.

In the directional spray systems, dust-laden air is moved along the face by air spray manifolds positioned between the drums (Figure 3-10). These sprays promote movement of dust-laden air along the face side of the shearer to prevent migration toward the walkway. Three or four manifolds containing three to five sprays each are typically spaced along the length of the shearer body. These manifolds are either located on the face side of the shearer or on the top of the shearer close to the face. All sprays are oriented downwind. Results from a series of underground tests showed that the shearer-clearer spray system reduced operator exposure from shearer-generated dust by about 50% when cutting against face ventilation and by at least 30% when cutting with ventilation [Ruggieri et al. 1983; Jayaraman et al. 1985].

Figure 3-10.—Directional sprays mounted on face side of shearer body.

- **Keeping the headgate splitter arm parallel to the top of the shearer.** Maintaining the position of the headgate splitter arm near parallel is critical to keeping dust from boiling out into the walkway, especially at higher-seam longwalls that are typically found at western longwall operations. During recent surveys [Rider and Colinet 2007], NIOSH personnel observed a hydraulically adjustable splitter arm that was angled down toward the pan line during head-to-tail passes, allowing respirable dust to migrate over the top of the splitter arm and into the walkway (Figure 3-11). Also, as mining advanced toward the headgate, NIOSH personnel noticed that a dust cloud would roll up under the splitter arm belting when the cutting drum was in the raised position and the splitter arm was angled upward (Figure 3-11). Positioning the splitter arm so that it is level with the shearer body and parallel to floor may prevent the dust cloud from migrating over or under the splitter arm and into the walkway.

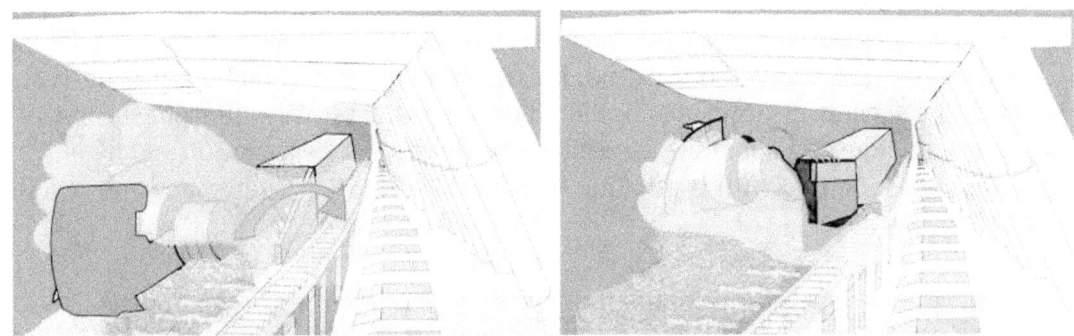

Figure 3-11.—Position of splitter arm may allow dust to migrate into walkway.

- **Shearer deflector plates.** The main function of the hydraulically controlled shearer deflector plates (Figure 3-12) is to protect shearer operators from debris flying off the face. In a raised position, the deflector plates seem to enhance the directional spray system effectiveness by providing a physical barrier that helps to confine contaminated air close to the face. The deflector plates should be raised as high as face conditions allow to provide maximum protection.

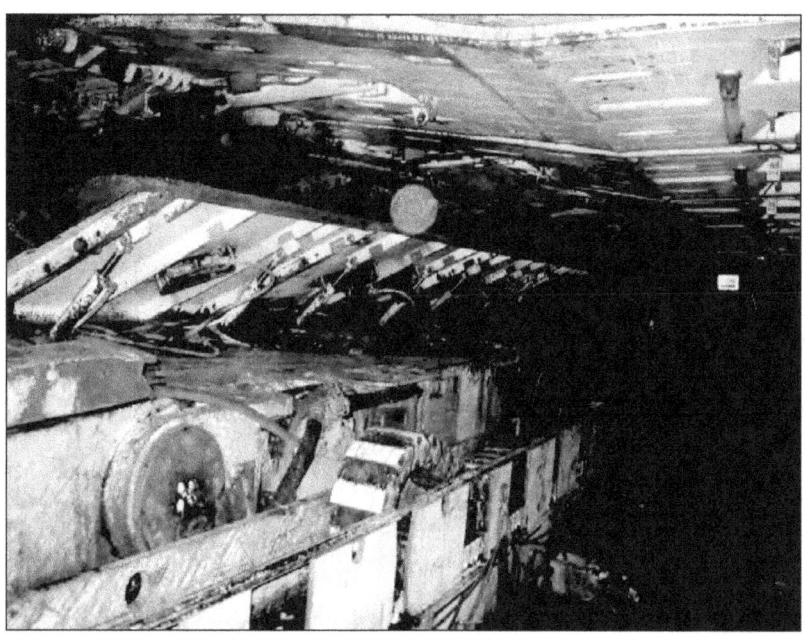

Figure 3-12.—Raised deflector plate can enhance the effectiveness of the directional spray system.

Shearer deflector plates have also been equipped with water sprays mounted in the plates, which can supplement the dust control effectiveness of the shearer-clearer system. However, shearer operators must be diligent in turning off the sprays if the deflector plate is lowered. If these sprays are operational when the deflector plate is down, the spray plume is directed upward and strikes the underside of the shields. This impact creates turbulence that can cause the ventilating airstream to carry dust out into the walkway, where it may adversely affect dust levels at and downwind of the shearer.

- **Crescent sprays.** Crescent sprays (Figure 3-13) can be located on each ranging arm and are typically oriented inward toward the cutting drum. These sprays are located on the top and end of the ranging arm. It is important that these sprays be aimed inward toward the cutting drum and appropriately spaced to provide uniform wetting of the entire cutting zone. Crescent sprays on the headgate ranging arm should be used with caution. Sprays on the end of the headgate ranging arm are oriented into the face airflow, which can create turbulence that forces dust toward the walkway [Colinet et al. 1997].

Figure 3-13.—Crescent sprays located on shearer ranging arm.

- **Lump breaker spray manifold.** Positioning a spray manifold at the end of the lump breaker and directing the spray down toward the conveyor can provide better, more uniform wetting of the cut coal. Using larger-orifice sprays operated at pressures less than 80 psi will provide higher volumes of water per spray wetting without creating turbulence.

- **Tailgate-side sprays.** Original directional spray systems were equipped with a splitter arm with sprays on the tailgate end of the shearer to help confine shearer-generated dust near the face. These splitter arm sprays also created a clean air envelope in the walkway downwind of the shearer, potentially reducing the dust exposure of the tailgate shearer operator and jack setters advancing shields near the shearer. Although use of the tailgate-side splitter arm has declined, a similar benefit was observed at mines that installed a spray manifold on the tailgate end of the shearer (Figure 3-14). These sprays are oriented parallel to the tailgate ranging arm or angled slightly toward the tailgate drum and act as a water curtain confining the dust cloud near the face. It is important that these sprays confine the dust along the face and not cause excessive turbulence that could cause the dust to migrate away from the cutting drum and into the walkway. These sprays may be able to carry water a distance of 10–20 ft downwind of the shearer if aligned properly and operated with sufficient flow and pressure. They can further enhance the air split created by the shearer's directional spray system.

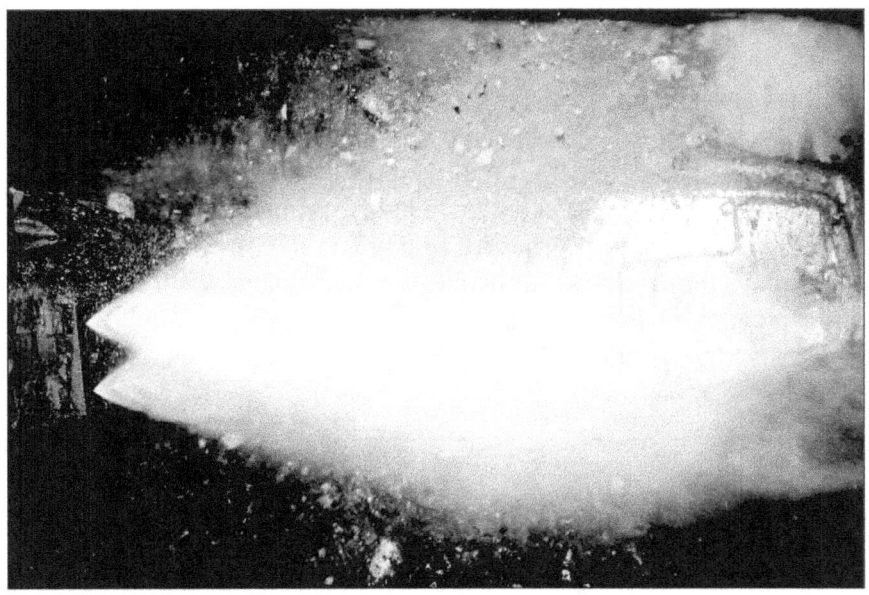

Figure 3-14.—Spray manifold mounted on tailgate end of shearer body.

CONTROLLING SHIELD DUST

Over the last several years, advances in longwall mining technology have resulted in more powerful and faster shearers capable of mining at cutting speeds exceeding 100 fpm. This also requires shields to advance at a faster rate. As shield supports are lowered and advanced, crushed coal and/or rock fall from the top of the shield canopy directly into the airstream ventilating the longwall face. Shield advance has become automated and is now initiated by the shearer position. Shields are typically being advanced within two or three shields of the trailing shearer drum. As a result, shield movement can be a significant source of dust exposure for shearer operators when shields are advanced upwind of the shearer during head-to-tail passes. Following is a discussion of observed spray systems that offer potential solutions for shield dust control.

- **Canopy-mounted spray systems.** Most of the dust liberated by shield movement comes from the canopy area of the shields during advance. A canopy spray system that activates sprays discharging into the roof material on top of the shields for a short period of time before and during shield advance has been available for many years. The goal is to wet the material on top of the canopy to lower dust levels during shield advance. Unfortunately, experience has shown that this type of system is hard to maintain and is not effective in distributing moisture to the material on top of the shield canopy.

- **Shield sprays on the underside of the canopy.** NIOSH researchers have observed shield sprays mounted on the underside of the shields, as shown in Figure 3-15 [Rider and Colinet 2007]. These sprays were automatically activated by the position of the shearer to create a moving water curtain in an attempt to contain the dust cloud near the headgate and tailgate drum areas. The location of these underside sprays ranged between the tip of each shield to an area above the spill plate. Each shield was equipped with one or two rows of two sprays. The sequencing of when the sprays were activated and deactivated was mine-specific. When the shield sprays were operational at one mine, researchers observed that they had a negative impact on controlling respirable dust associated with the upwind drum. The shield sprays interacted with the upwind splitter arm sprays, creating turbulence that resulted in a dust and mist cloud rolling into the walkway. Proper on/off sequencing of these shield sprays is critical for these sprays to supplement the directional spray system. Properly aligned sprays directed toward the face with sufficient water pressure and volume have the potential to enhance the envelope of clean air created by the shearer's directional spray system.

Figure 3-15.—Shield sprays located on the underside of the canopy.

- **Air dilution.** In theory, supplying additional air to the face should increase the dilution of dust liberated by shield advance. However, to increase air quantity coming onto the face, the velocity of the airstream must be increased. Increasing air velocity could provide greater potential for dust entrainment because the relatively dry shield dust falls directly into the airstream. Thus, the amount of shield dust that must be controlled at the shearer may be significantly higher, especially in recent years where automated shield advance is occurring within a few shields of the trailing shearer drum. If shields are advanced upwind of the shearer, it should be done as far upwind as possible without creating operational problems. This may allow dust generated by the shield movement to mix with clean air and dilute before it reaches the shearer operators.

- **Unidirectional cutting sequence.** Unidirectional cutting may allow for greater flexibility to place workers upstream of the dust sources than bidirectional cutting [Kissell et al. 2003]. Depending on roof conditions, this may allow the operators to modify the cut sequence so that shields are only advanced downwind of the shearer. Activating shield advance as close to the tailgate drum as possible and keeping jack setters upwind of the advancing shields may protect the jack setters from elevated dust levels by keeping them in a clean air envelope created by the shearer's directional spray system.

ALTERNATE DUST CONTROL TECHNOLOGIES

- **Ventilated cutting drums.** About 50% of the dust generated on the longwall comes from the cutting action of the shearer drums. It is known that once the respirable dust becomes airborne, it is difficult to control and is best reduced by capture and suppression at the source. Research studies [Fench 1983; Divers et al. 1987] have shown that a ventilated cutting drum (Figure 3-16) is effective in lowering respirable dust levels of the shearer operators. The ventilated drum is designed to reduce the amount of dust from the cutting zone through 12 water-powered dust capture tubes built into the hub of the shearer drum. High-pressure water is released from a spray ring manifold on the face side of the drum. The sprays act as a fan and scrubber to induce dust-laden air from the face side of the drum through tubes. The tubes are open-ended and contain hollow-cone water sprays. At the gob side of the drum, a deflector plate attached to the cowl arm deflects the water spray away from the operator. High-pressure water at about 1,000 psi is required for maximum air movement, efficient dust collection, and for cleaning the tubes.

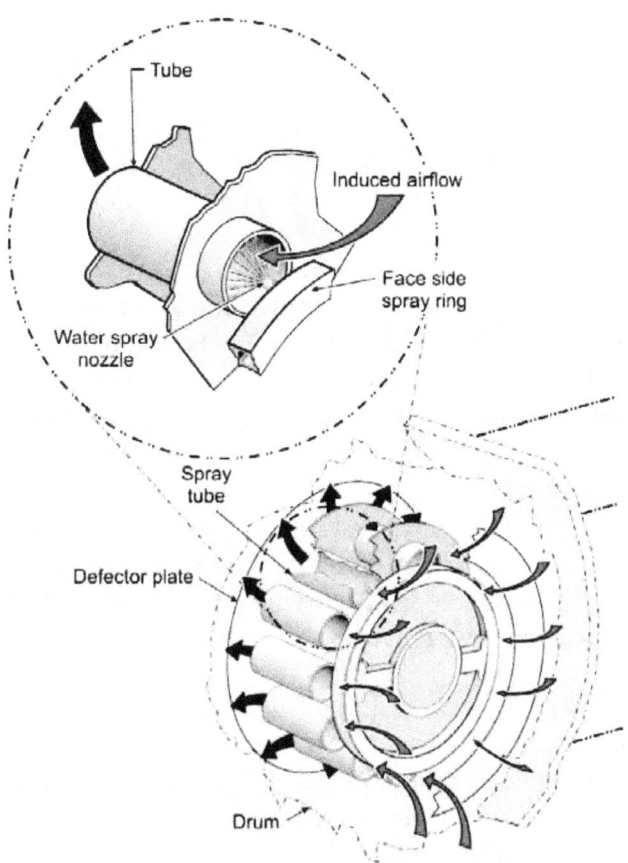

Figure 3-16.—Schematic of ventilated shearer drum.

The use of ventilated cutting drums [Divers et al. 1987] reduced dust levels by about 50%. Maintenance issues and significant capital investment are major concerns with the system. Maintaining a high-pressure water supply system on the longwall is also a significant concern, along with the need to custom design and build a specialized hub for each drum. This technology may not be applicable on drums less than 52 inches in diameter because of space needed for the vanes and shrouded sprays. Although design and operational constraints were not overcome at the time of this research, significant dust level reductions were achieved. This system may have merit with modifications through technological advances that have taken place since this research was conducted in the 1980s.

- **Foam discharge from shearer drum.** The discharge of foam from large-diameter nozzles located in the shearer drum has shown promise in reducing the shearer operator's respirable dust exposures. A U.S. Bureau of Mines research study [Laurito and Singh 1987] showed that foam can distribute a given quantity of moisture much more evenly over a large surface area. However, a high degree of mixing must take place between the foam and the cut coal for the foam application to be successful in reducing dust levels. Also, any chemical additive such as foam or wetting agents have the potential to disrupt the coal cleaning in the preparation plant depending on the type of cleaning used at the plant. This must be considered when selecting a water additive for dust control. As stated earlier, the optimum location to attack dust generated by the cutting action of the drums is at the source. In the study by Laurito and Singh [1987], a compressed-air foam generation system that discharged foam through 10–12 nozzles on the shearer drum was evaluated at longwall operations. Results from the tests showed that shearer operator dust exposures were reduced by 50%–70%. Like the ventilated drum, a foam distribution system adds complexity, maintenance, and cost, but offers the potential for improved dust control of shearer generated dust.

- **High-pressure inward-facing drum sprays.** High-pressure (up to 1,200 psi), inward-facing drum sprays [Jankowski et al. 1989] have been shown to confine the dust generated by the cutting drum to the face area. This spray system seemed to suppress the amount of dust that becomes airborne by improving the moisture distribution efficiency of the drum sprays. The high-pressure, inward-facing drum water spray system consists of high-pressure sprays located in each bit block and oriented toward the coal face. Based on field tests at a longwall operation [Jankowski et al. 1989], high-pressure, inward-facing drum sprays were most effective at reducing dust levels with the sprays oriented at 30° and operated at 800 psi. Dust exposure levels under these conditions were reduced by 39%.

Testing has shown that the use of high-pressure, inward-facing drum sprays had some constraints. The most objectionable from an operation standpoint was clogging of the sprays. Rust particles from the drum were a potential source of water contamination, increasing the risk of the sprays clogging. To minimize clogging, the spray system water supply should be filtered. With some refinements and improvements, this system may help keep dust generated by cutting drums confined to the face area.

REFERENCES

Baig NA, Dean AT, Skiver DW [1994]. Successful use of belt washers. In: Proceedings of the American Power Conference. Chicago, IL. Illinois Institute of Technology, pp. 976–978.

Breuer H [1972]. Progress in dust and silicosis control. Glückauf *108*(18):806–814.

Chekan GJ, Listak JM, Colinet JF [2001]. Laboratory testing to quantify dust entrainment during shield advance. In: Proceedings of the Seventh International Mine Ventilation Congress (Krakow, Poland, June 17–22, 2001), pp. 291–298.

Chekan GJ, Listak JM, Colinet JF [2004]. Factors impacting respirable dust entrainment and dilution in high-velocity airstreams. In: Yernberg WR, ed. Transactions of Society for Mining, Metallurgy, and Exploration, Inc. Vol. 316. Littleton, CO: Society for Mining, Metallurgy, and Exploration, Inc., pp. 186–192.

Colinet JF, Spencer ER, Jankowski RA [1997]. Status of dust control technology on U.S. longwalls. In: Ramani RV, ed. Proceedings of the Sixth International Mine Ventilation Congress. Chapter 55. Littleton, CO: Society for Mining, Metallurgy, and Exploration, Inc., pp. 345–351.

Divers EF, Jankowski RA, Kelly J [1987]. Ventilated drum controls longwall dust and methane. In: Proceedings of the Third U.S. Mine Ventilation Symposium (October 12–14, 1987), pp. 85–89.

EIA [2009]. Annual coal report, 2007. Washington DC: U.S. Department of Energy, Energy Information Administration, DOE/EIA-0584 (2007), p. 17.

French AG [1983]. The extraction of respirable dust from machines working on longwall faces. National Coal Board, Mining Research and Development Establishment, Proceedings of the European Economic Communities Conference on Dust Control, Luxembourg, pp. 57–93.

Jankowski RA, Colinet JF [2000]. Update on face ventilation research for improved longwall dust control. Min Eng *52*(3):45–52.

Jankowski RA, Kissell FN, Daniel JH [1986]. Longwall dust control: an overview of progress in recent years. Min Eng *28*(10):953–958.

Jankowski RA, Whitehead KL, Thomas DJ, Williamson AL [1989]. High-pressure inward-facing drum sprays reduce dust levels on longwall mining sections. In: Proceedings of Longwall USA (Pittsburgh, PA), pp. 231–242.

Jayaraman NI, Kissell FN, Cross W, Janosik J, Odoski J [1981]. High-pressure shrouded water sprays for dust control. Pittsburgh, PA: U.S. Department of the Interior, Bureau of Mines, RI 8536. NTIS No. PB 81-231458.

Jayaraman NI, Jankowski RA, Kissell FN [1985]. Improved shearer-clearer system for double-drum shearers on longwall faces. Pittsburgh, PA: U.S. Department of the Interior, Bureau of Mines, RI 8963. NTIS No. PB 86-107844.

Kelly J, Ruggieri S [1990]. Evaluate fundamental approaches to longwall dust control; subprogram C: Stageloader dust control. Foster-Miller, Inc. U.S. Bureau of Mines contract J0318097. NTIS No. DE 90-015510.

Kissell FN, Stachulak JS [2003]. Underground hard-rock dust control. In: Kissell FN, ed. Handbook for dust control in mining. Pittsburgh, PA: U.S. Department of Health and Human Services, Centers for Disease Control and Prevention, National Institute for Occupational Safety and Health, DHHS (NIOSH) Publication No. 2003-147, IC 9465, pp. 83–96.

Kissell FN, Colinet JF, Organiscak JA [2003]. Longwall dust control. In: Kissell FN, ed. Handbook for dust control in mining. Pittsburgh, PA: U.S. Department of Health and Human

Services, Centers for Disease Control and Prevention, National Institute for Occupational Safety and Health, DHHS (NIOSH) Publication No. 2003-147, IC 9465, pp. 39–55.

Kost JA, Yingling JC, Mondics BJ [1981]. Guidebook for dust control in underground mining. Bituminous Coal Research Inc. U.S. Bureau of Mines contract J0199046. NTIS No. PB 83-109207.

Laurito AW, Singh MM [1987]. Evaluation of air sprays and unique foam application methods for longwall dust control. Engineers International, Inc. U.S. Bureau of Mines contract J0318095. NTIS No. PB89-189922.

Listak JM, Chekan GJ, Colinet JF [2001]. Laboratory evaluation of shield dust entrainment in high velocity airstreams. In: Transactions of Society for Mining, Metallurgy, and Exploration, Inc. Vol. 310. Littleton, CO: Society for Mining, Metallurgy, and Exploration, Inc., pp. 155–160.

MSHA [1989]. Belt entry ventilation review: report of findings and recommendations. Arlington, VA: U.S. Department of Labor, Mine Safety and Health Administration.

MSHA [2009]. Program Evaluation and Information Resources, Standardized Information System. Arlington, VA: U.S. Department of Labor, Mine Safety and Health Administration.

Mundell RL et al. [1979]. Respirable dust control on longwall mining operations in the United States. In: Proceedings of the Second International Mine Ventilation Congress (Reno, NV, November 4–8, 1979).

Niewiadomski GE [2009]. Mine Safety and Health Administration, private communication.

NIOSH [2008]. Work-related lung disease surveillance report, 2007. Morgantown, WV: U.S. Department of Health and Human Services, Centers for Disease Control and Prevention, National Institute for Occupational Safety and Health, DHHS (NIOSH) Publication No. 2008-143a.

Organiscak JA, Reed WR [2004]. Characteristics of fugitive dust generated from unpaved mine haulage roads. Int J Surface Min Reclam Environ *18*(4):236–252.

Organiscak JA, Jankowski RA, Kelly JS [1986]. Dust controls to improve quality of longwall intake air. Pittsburgh, PA: U.S. Department of the Interior, Bureau of Mines, IC 9114. NTIS No. PB 87-167573.

Organiscak JA, Khair AW, Ahmad M [1996]. Studies of bit wear and respirable dust generation. In: Transactions of Society for Mining, Metallurgy, and Exploration, Inc. Vol. 298. Littleton, CO: Society for Mining, Metallurgy, and Exploration, Inc., pp. 1932–1935.

Organiscak JA, Page SJ, Cecala AB, Kissell FN [2003]. Surface mine dust control. In: Kissell FN, ed. Handbook for dust control in mining. Pittsburgh, PA: U.S. Department of Health and Human Services, Centers for Disease Control and Prevention, National Institute for Occupational Safety and Health, DHHS (NIOSH) Publication No. 2003-147, IC 9465, pp. 73–81.

Potts JD, Jankowski RA [1992]. Dust considerations when using belt entry air to ventilate work areas. Pittsburgh, PA: U.S. Department of the Interior, Bureau of Mines, RI 9426.

Rider JP, Colinet JF [2007]. Current dust control practices on U.S. longwalls. In: Proceedings of Longwall USA (Pittsburgh, PA, June 5–7, 2007).

Ruggieri SK, Muldoon TL, Schroeder W, Babbitt C, Rajan S [1983]. Optimizing water sprays for dust control on longwall shearer faces. Foster-Miller, Inc. U.S. Bureau of Mines contract J0308019. NTIS No. PB 86-205408.

Shirey CA, Colinet JF, Kost JA [1985]. Dust control handbook for longwall mining operations. BCR National Laboratory. U.S. Bureau of Mines contract J0348000. NTIS No. PB-86-178159/AS.

Stahura RP [1987]. Conveyor belt washing: Is this the ultimate solution? TIZ-Fachberichte *111*(11):768–771.

Tomb TF et al. [1992]. Evaluation of respirable dust control on longwall mining operations. In: Transactions of Society for Mining, Metallurgy, and Exploration, Inc. Vol. 288. Littleton, CO: Society for Mining, Metallurgy, and Exploration, Inc., pp. 1874–1878.

USBM [1985]. Technology news 224: Improved stageloader dust control in longwall mining operations. Pittsburgh, PA: U.S. Department of the Interior, Bureau of Mines.

CHAPTER 4.—CONTROLLING RESPIRABLE DUST ON CONTINUOUS MINING OPERATIONS

By Jeffrey M. Listak[1]

This chapter discusses proven methods and engineering controls to minimize respirable dust concentrations on continuous mining operations. The highest respirable dust concentrations on continuous mining sections are generated from two sources: the continuous miner and the roof bolter. Also, continuous miner and roof bolter operators are often exposed to elevated silica levels as a result of cutting or drilling into rock.

The occupation on a mechanized mining unit with the highest dust exposure, based on results of respirable dust samples collected by MSHA, is classified as the designated occupation (DO) by MSHA. In addition to being sampled by MSHA, each DO is also sampled bimonthly by the mine operator. The samples are then submitted to MSHA for analysis to determine compliance with the applicable dust standard. MSHA lowers the dust standard below 2 mg/m^3 if the silica content of the sample exceeds 5% by weight in dust samples. For operations on reduced dust standards, MSHA inspector samples from 2004–2008 show that 20% of miner operator samples and 10% of bolter operator samples exceeded their applicable reduced dust standard [MSHA 2009].

In this chapter, controls for continuous miners and roof bolters are discussed in detail. Controls for other related sources of dust, such as intake air, are also described.

CONTINUOUS MINER DUST CONTROL

The greatest source of respirable dust at continuous mining operations is the continuous mining machine. At most continuous mining operations, the DO is the continuous miner operator. Dust generated by the continuous miner has the potential to expose the miner operator and anyone working downwind of the active mining.

As with any dust source, air and water are used to dilute, suppress, redirect, or capture dust. Ventilating air to a continuous mining section, whether blowing or exhausting, is the primary means of protecting workers from overexposure to respirable dust. Proper application of water spray systems, ventilation, and mechanical equipment (scrubbers) provides the best overall means of respirable dust control. Maintenance of scrubbers, water sprays, and bits are basic to any effective dust control strategy and must be routinely practiced. Suppression of dust is the most effective means of dust control. Suppression is achieved by the direct application of water, usually at the point of attack, to wet the coal before and as it is broken to prevent dust from becoming airborne. Once dust is airborne, other methods of control must be applied to dilute it, direct it away from workers, or remove it from the work environment. Redirection of dust is achieved by water sprays that move dust-laden air in a direction away from the operator and into the return entry or behind the return curtain. Capture of dust is achieved either by water sprays that impact with the dust in the air to remove it or by mechanical means (e.g., fan-powered dust collectors). Ventilating air dilutes and directs dust away from workers. Either blowing or

---

[1]Mining engineer, Office of Mine Safety and Health Research, National Institute for Occupational Safety and Health, Pittsburgh, PA.

exhausting ventilation is used on continuous mining sections. Advantages and disadvantages of each method will be described.

Water Spray Systems

There are several types of water sprays available for use on continuous miners to control dust. Spray nozzle type, location, pattern, flow, and pressure are all factors to consider when designing a spray system. The type of spray used at a particular location depends on the desired application. For example, for suppression of dust, high flow at low pressure close to the source is most effective. For airborne dust capture, smaller high-velocity droplets are required to impact with dust and remove it from the air. For redirection, higher pressure is required. Figure 4-1 shows the sprays most commonly used for controlling dust. A description of the spray patterns used and their applications is also provided.

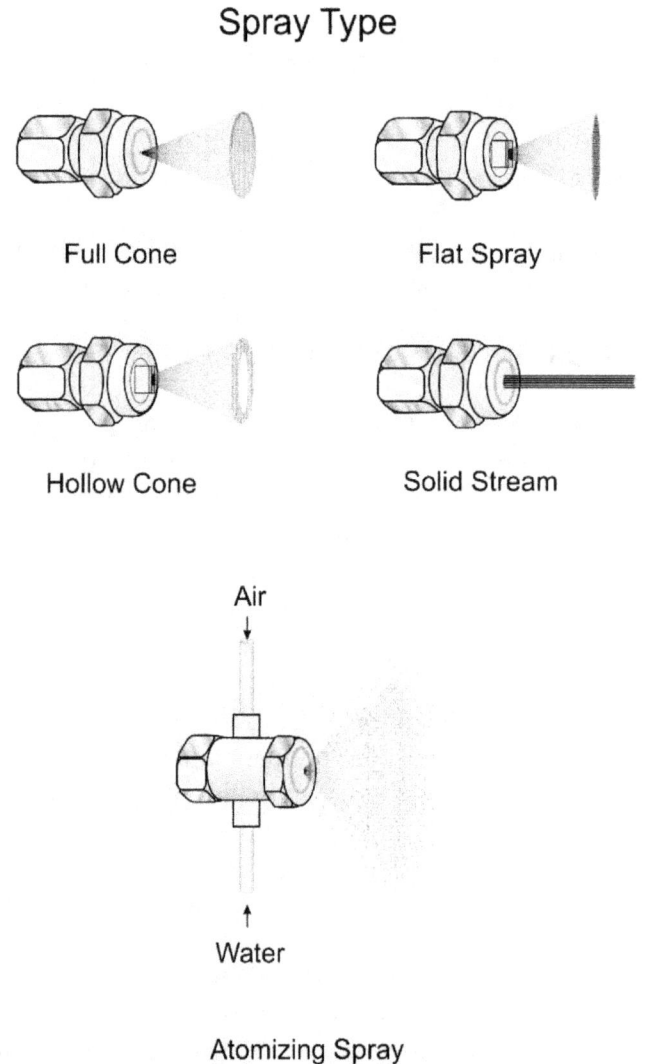

Figure 4-1.—Spray types used for dust control in mining.

- **Hollow-cone.** Hollow-cone sprays use a circular, outer-ring spray pattern in three different designs: whirl chamber, deflected, and spiral sprays. These sprays produce small to medium droplet sizes. Hollow-cone sprays are the best choice for most practical mining applications because they have larger-orifice nozzles and are less likely to clog [Kissell 2003]. Standard water spray systems from the manufacturer typically locate hollow-cone sprays on the boom directly behind the cutter head, on the underside of the boom, and along the sides of the cutter head. They are particularly effective for dust knockdown, as well as redirecting dust away from the worker.

- **Full-cone.** Full-cone sprays use a solid cone-shaped spray pattern with a round impact area that provides high velocity over a distance. They produce medium to large droplet sizes over a wide range of pressures and flows. They are normally used when the sprays need to be located farther away from the dust source or when a uniform wetting pattern is desired, such as scrubber filters or belt transfer points. Full-cone sprays are also useful to wet the throat area of the miner's conveyor to suppress dust during transfer to shuttle cars.

- **Solid-stream.** A straight, solid, uniform stream of water at high flow and low pressure is the goal of solid-stream sprays. They provide uniformity of wetting of the material to be cut. They are designed to be used close to the source to deluge the area before cutting or loading. The objective of solid-stream sprays is dust suppression.

- **Flat-fan.** The flat-fan spray pattern comes in three different designs: tapered, even, and deflected-type sprays. Flat-fan sprays produce small to medium droplet sizes over a wide range of flows and spray angles and are normally located in narrow, enclosed spaces. Locating flat-fan sprays along the side of the cutter head helps to contain dust under the boom, allowing for capture by the scrubber inlets. Flat-fan sprays are effective for dust containment.

- **Air-atomizing.** Air-atomizing spray patterns are available in two different designs: hydraulic and air-assisted. Hydraulic nozzles produce fine-mist droplet sizes and have low volume capacities. Air-assisted nozzles produce the smallest droplets of all sprays, but are the most expensive and complex to install because they require compressed air.

  Figure 4-2 shows the airborne capture performance of the different spray nozzles at different operating pressures and compares the relative effectiveness of each spray type. Although air-atomizing nozzles produce the best airborne dust capture, their use in mines is impractical due to high maintenance requirements (i.e., they are prone to clogging) and the need to supply compressed air to each nozzle.

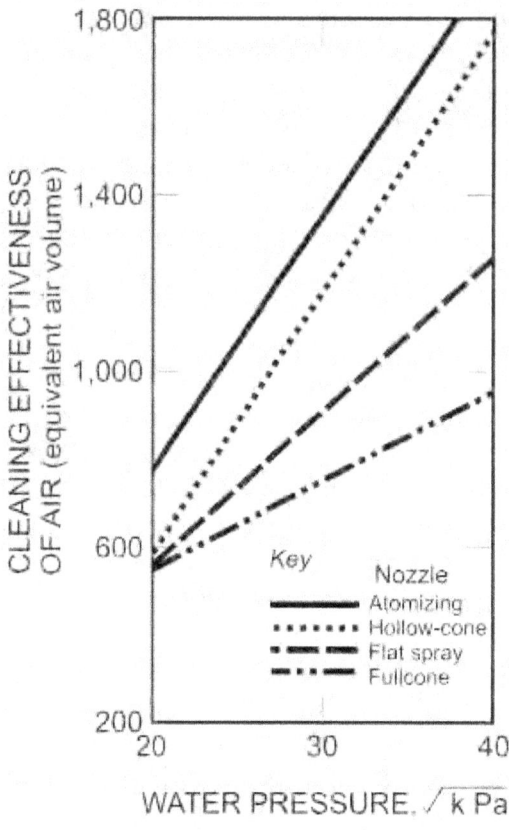

Figure 4-2.—Relative spray effectiveness of four spray nozzles used in mining.

Most continuous miners use a combination of spray types to achieve the best control. Although higher water pressure will raise the effectiveness of water sprays, a marked disadvantage is that it entrains large volumes of air and subsequently dust. This can result in dust rollback. The earliest water sprays on a continuous miner were used for bit lubrication, bit cooling, and dust control. Although these sprays controlled respirable dust exposure to a limited extent, they also created large quantities of turbulence and dust rollback. Dust rollback over the continuous miner infiltrated the operator's position, resulting in dust overexposure. To control rollback, sprays were relocated atop and beneath the cutting drum. The top sprays operated at a pressure of 100 psi and a flow rate of 0.95 gpm per spray. Two large-orifice, deluge-type sprays were mounted on the left and right underside of the boom and directed to spray into the cutting bits. These sprays operated at a low pressure of about 7 psi and a higher flow rate of 5 gpm per spray. Dust rollback was decreased as a result because the spray droplets moved only a short distance before impacting on the cutting bits. The short distance also increased coal surface wetting capabilities while minimizing turbulence (Figure 4-3). In-mine evaluations of these boom sprays show that miner operator dust exposures are reduced by 40% compared to the factory-installed spray system [Schroeder et al. 1986].

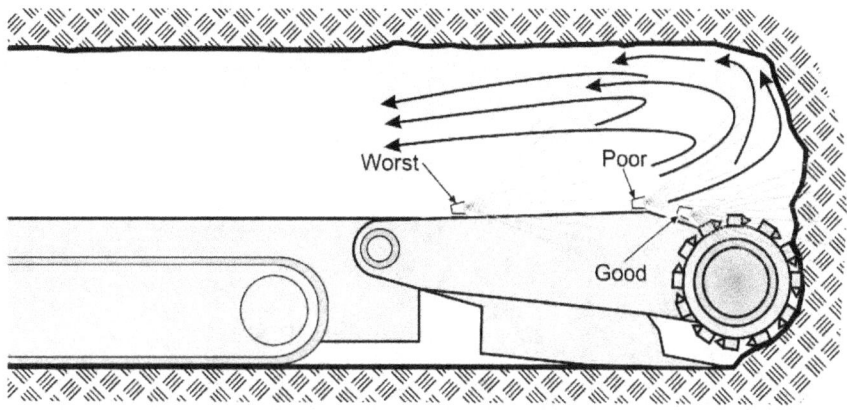

Figure 4-3.—Spray location impact on dust rollback.

High-pressure sprays, installed at the rear corner of the shovel on the side opposite from the exhausting ventilation curtain, can sweep underboom dust toward the curtain. Extensive underground testing showed that the shovel sprays reduced coal dust exposures by 60% at the miner operator's location while virtually eliminating exposures to respirable quartz dust [Schroeder et al. 1986].

The following practices have been shown to reduce dust exposures on continuous mining operations:

- Dust rollback over the miner can be caused by high-pressure (>100 psi), wide-angle cone sprays [Jayaraman et al. 1984]. A typical miner spray does most of its airborne dust collection in the first 12 in. Thus, top and side nozzles should be arranged for "low" reach and no overspray (Figure 4-4, A and B). Past research has shown that flat-fan sprays at a horizontal orientation with high flow and low pressure (<100 psi) across the boom, located as close to the cutter head as possible, provide uniform coal wetting across the cutter head during mining while limiting rollback. Large-orifice, low-pressure deluge throat sprays should be used under the boom on flight conveyors at 5 gpm. Broken material should be wetted as it is gathered and conveyed (Figure 4-4, C) [Schroeder et al. 1986].

- For dust containment at the face, flat-fan sprays, located 1 ft back from the cutter head on both sides of the miner with a vertical spray pattern that is oriented 30° from the miner body, act as blocking sprays and will help contain dust, thus enabling improved dust capture by the scrubber inlets [Goodman 2000]. Laboratory testing has shown that increasing spray nozzle pressure and/or the width of the spray angle will increase the airflow induced by that nozzle, potentially increasing its effectiveness as a blocking spray [Pollock and Organiscak 2007].

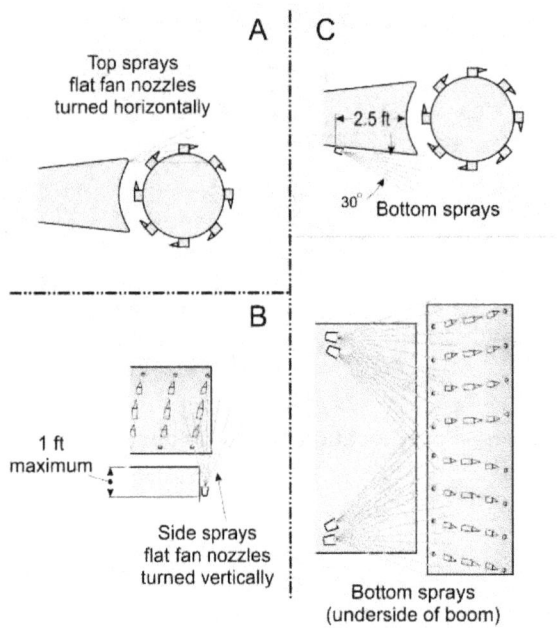

Figure 4-4.—Antirollback spray system for miner.

- High-pressure sprays are recommended for redirecting of dust. High pressure (>150 psi) raises the efficiency per unit of water [Jayaraman and Jankowski 1988] and is effective for air moving (Figure 4-5). However, care must be taken when determining location and direction because high pressure can cause turbulence, leading to rollback.

- A directional spray system design (spray fan) is a water-powered ventilation system originally designed to sweep methane gas toward the return. This system contains several spray manifolds placed on the continuous mining machine to direct fresh intake air to the cutting face, sweep contaminated air across the face, and direct this airflow into the return airway. In practice, the spray fan system is used only with an exhaust face ventilation scheme. The spray fan design has been successful for methane control at the face, but is not as effective for dust control [Goodman et al. 2004].

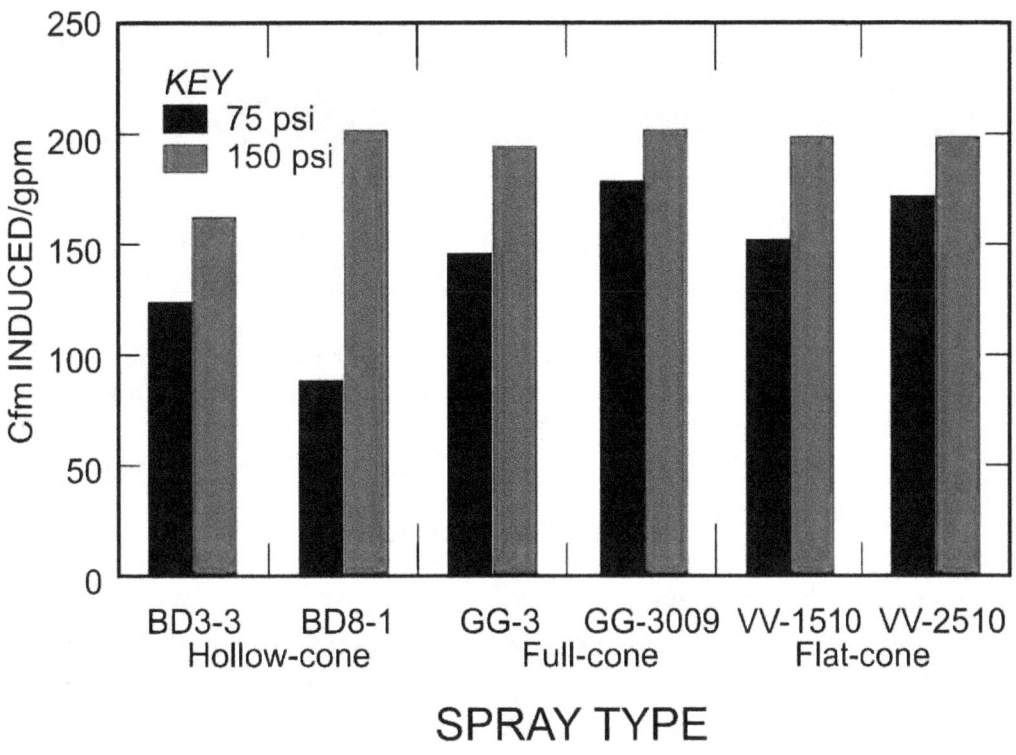

Figure 4-5.—Air-moving effectiveness of different spray types.

- Recently developed wet-head continuous miners have water sprays mounted on the cutter head directly behind the cutter bits. Water sprays in this position cool bits, which reduce frictional ignitions and have the potential for reducing dust generation during cutting. Available spray nozzles are either solid-cone spray patterns with a 1-mm orifice and flow of 0.4 gpm at 100 psi, or hollow-cone patterns with a flow of 0.2 gpm at 100 psi. However, wet-head technology has not yet consistently demonstrated dust reduction benefits in reported studies. These studies have shown variation in dust reductions under its current configuration [Strebig 1975; Goodman et al. 2006]. One benefit that has been reported by mining machine operators is increased visibility at the cutter head when the wet head is used. This may lead to better control of the cutting head and have a beneficial impact on required maintenance (reduced bit changes) and subsequent dust generation.

- Good water filtration greatly aids in spray system effectiveness. Dirt and rust particles in the water line can cause frequent clogging of spray nozzles. A simple, nonclogging water filtration system is available and should be used to replace conventional spray filters [Divers 1976].

- Operators should examine, clean, and/or replace sprays if necessary before each cut.

- A cut sequence should be adopted so that cut-throughs are made from intake to returns when practical to prevent return air from blowing back over the operator [Fields et al. 1991].

- Handheld remote control of the continuous miner has made it possible for operators to stay outby the miner while operating the machine. However, operator positioning is crucial depending on the ventilation scheme being used. Correct positioning will be discussed later in the "Face Ventilation" section.

- Remote control also enables mines to advance farther than the traditional 20-ft cut, if approved by MSHA. However, deep-cut mining requires that additional ventilation and dust control measures be incorporated. The machine-mounted scrubber is a critical component for deep-cut mining [Schultz and Fields 1999].

Flooded-Bed Scrubbers

Remotely controlled continuous miners allow the operator to remain under supported roof while the miner can advance to cut depths up to 40 ft, if approved by MSHA. Extended cuts reduce the number of face changes required of the miner, which can lead to higher production. Therefore, most U.S. continuous mining operations are taking deep or extended cuts. However, remote control operation does not allow the operator to advance the ventilation curtain to dilute the face of dust and methane. As a result, deep-cut mining has benefited from the installation of fan-powered, flooded-bed scrubbers on miners. Scrubber inlets mounted close to the cutter head help move air toward the face and capture dust near the face.

Flooded-bed scrubbers capture dust-laden air from the cutting face, carry this air through ductwork on the miner, and pass the air through a filter panel that is wetted with water sprays (Figure 4-6). As dust particles impact and travel through the filter panel, they mix with water droplets and are removed from the airstream by a mist eliminator. The cleaned air is discharged from the scrubber back into the mine environment. The density and type of media used in the filter panel influence the dust collection efficiency and air-moving capacity of the scrubber. Optimum flooded-bed scrubber performance is achieved when all of the dust-laden air at the cutting face is drawn into the scrubber and a high percentage (>90%) of the respirable dust is removed from this air [NIOSH 1997].

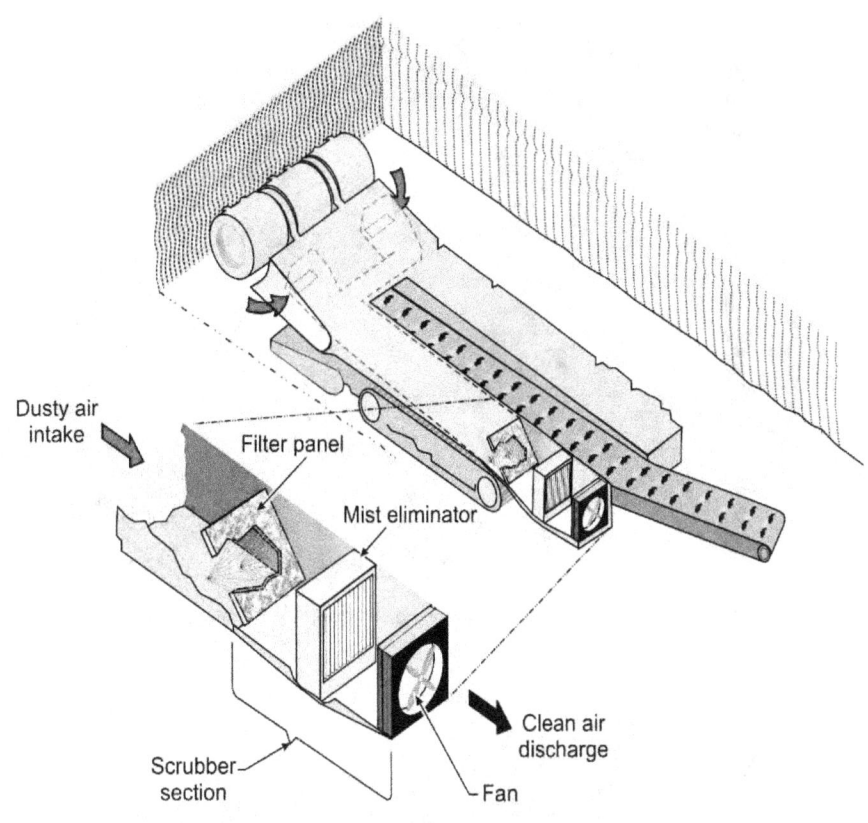

Figure 4-6.—Components and design of a flooded-bed scrubber.

The following practices have been shown to improve the efficiency of the scrubber:

- **Scrubber maintenance.** Scrubbers lose as much as one-third of their airflow after just one cut [Schultz and Fields 1999]. The most common cause of efficiency loss is filter panel clogging. Pitot tubes should be used to obtain air velocity readings as a measure of scrubber performance. When the dust is excessive, cleaning of the filter panel (Figure 4-7), the demister (Figure 4-8), and the scrubber ductwork is required more often. Also, the spray nozzles in the ductwork should be checked to ensure they are completely wetting the entire filter panel and not just the center.

Figure 4-7.—Cleaning scrubber filter panel with water spray.

Figure 4-8.—Cleaning the demister with a water nozzle.

One major manufacturer recommends the following cleaning schedule for a flooded-bed dust collector:

(1) *Twice each shift:* Clean filter with water.
(2) *Once each shift:* Replace filter with cleaned filter. Back-flush dirty filter with water and allow to dry. When dry, shake remaining dirt from filter before reusing.
(3) *Daily:* Wash inlets and ductwork with water.
(4) *Weekly:* Wash venturi, sump, and demister module.

This recommended maintenance schedule is for general operation. However, field investigations have shown that in some instances more frequent filter cleaning is necessary. In some mines, filters should be cleaned with water at least after each place change. In addition, inlets and ductwork may require more frequent cleaning. The operator's approved mine ventilation plan will specify the specific maintenance schedule to be followed.

- **Airflow measurement.** MSHA requires a minimum airflow of 3,000 cfm to ventilate the active face (30 CFR[2] 75.325). However, when scrubbers are used, MSHA typically recommends that the face airflow be at or slightly above the airflow capacity of the scrubber. Consequently, MSHA periodically requires pitot tube traverse measurements of airflow through the scrubber. This scrubber airflow will then be considered when MSHA sets the minimum face airflow that is required. MSHA recommends that this face airflow be available and measured with the scrubber turned off. In practice, most operations are supplying quantities above the statutory minimum of 3,000 cfm in an effort to better control respirable dust and methane.

- **Filter panel thickness.** The thickness of the filter panel controls the filter collection efficiency. Of the 10-, 20-, and 30-layered panels available, the 30-layer panel is the most efficient (>90%) in capturing respirable-sized dust [NIOSH 1997] (Figure 4-9). However, it should be noted that the thicker filter panels will increase pressure drop and reduce the quantity of airflow through the scrubber. Ensure that the filter sprays are working properly and provide complete coverage of the filter media. The filter spray is typically a low-pressure (<50 psi), full-cone nozzle spray.

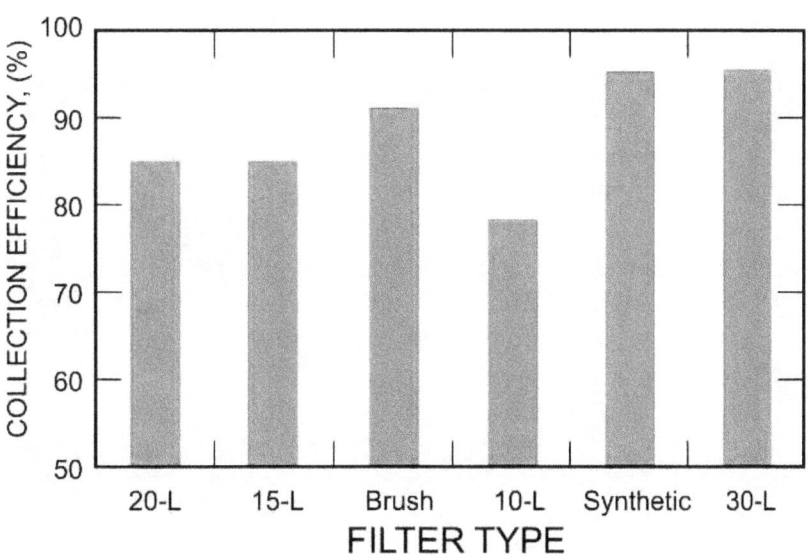

Figure 4-9.—Dust collection efficiency of scrubber filter panels.

---

[2]Code of Federal Regulations. See CFR in references.

- **Dust capture factors.** Overall performance of a flooded-bed scrubber depends on the collection efficiency of the filter panel and the capture efficiency, which is the amount of ventilating air drawn into the unit [Colinet and Jankowski 2000]. The machine design factors that impact capture efficiency are the scrubber air quantity and the location of the inlets. The air quantity should always be as large as possible and the inlets as far forward and close to the cutting drum as practical. Increasing filter density improves silica collection but also reduces the quantity of air that is drawn through the scrubber [Jayaraman et al. 1992]. In addition, the scrubber should continue to operate for 10–12 sec after coal cutting ends to allow the scrubber to capture dust remaining in the face.

- **Use of surfactants.** Surfactants have the potential to increase the wettability of dust by reducing the surface tension of the water and improving the capture of dust particles. In one trial, the use of surfactants in the scrubber sprays at a concentration of 0.013% by weight showed dust reductions as high as 31% [Hirschi et al. 2002].

- **Redirected scrubber discharge.** Face ventilation on sections with low mean entry air velocity can be augmented by redirecting scrubber discharge toward the face. However, the application of redirected scrubber air depends on the amount of methane being liberated at the face. A preliminary NIOSH study showed that redirecting a portion of the scrubber discharge toward the face is successful at reducing dust levels in the face entry. NIOSH also completed a brief study of scrubber redirection at an eastern mine that showed this method reduced dust exposures of the shuttle car operators by over 50%. Its impact on miner operator exposures was somewhat less, showing the need for maintaining proper balance between the redirected scrubber flow and face ventilation flow.

Bit Type and Wear

Bit type and bit wear can adversely affect respirable dust concentrations. Routine inspection of bits and replacement of dull, broken, or missing bits improve cutting efficiency and help minimize dust generation. Also, a study showed that bits designed with large carbide inserts and smooth transitions between the carbide and steel shank typically produce less dust [Organiscak et al. 1996] (see Figure 4-10). Lab studies on conical cutting bits have shown that significantly worn bits without their carbide tips produce much more dust [Organiscak et al. 1996].

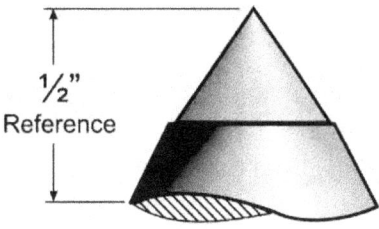

- Large carbide tip
- Smooth carbide to steel transition
- Low wear rate,
- Low dust levels

Figure 4-10.—Proper bit design can lower dust generation.

Modified Cutting Method

If roof rock must be cut, it is often beneficial to cut the coal beneath the rock first and then back the miner up to cut the remaining rock. This method of cutting leaves the rock in place until it can be cut out to a free, unconfined space, which creates less respirable dust (especially silica dust) [Jayaraman et al. 1988]. Figure 4-11 shows the modified cutting cycle.

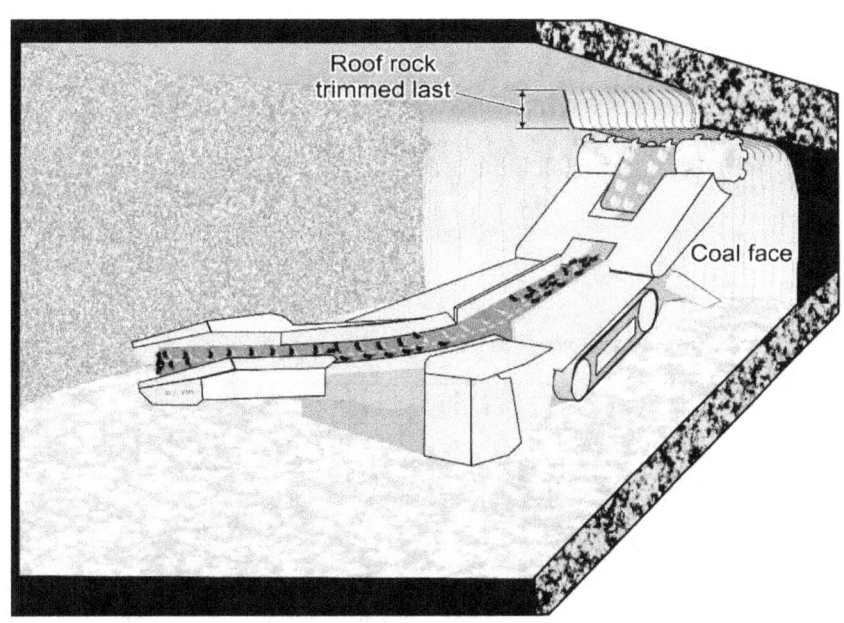

Figure 4-11.—Modified cutting cycle can lower dust generation.

FACE VENTILATION

The velocity and quantity of face ventilating air are important factors for controlling respirable dust exposure of the continuous miner operator. A good ventilation plan consists of sufficient mean entry air velocity to confine dust near the face and/or direct it toward the return entry with high enough quantity for diluting generated respirable dust. The two ventilation schemes most widely used for underground coal mining are blowing and exhausting. There are advantages and disadvantages to both systems as they relate to face worker dust exposure.

Blowing Face Ventilation

When blowing ventilation is used, intake air is delivered to the face of the working entry by blowing it from behind line brattice or tubing. The clean air is blown toward the face and sweeps the dust-laden air toward the return entries. This system allows the continuous miner operator to be positioned in the clean discharge air at the end of the blowing curtain or tubing. Although this method effectively sweeps dust and methane from the face, it also positions mobile equipment operators (e.g., shuttle car operators) and roof bolter operators working downwind in return air. Continuous miner operator movement is restricted due to the need to be positioned in the discharge air at the end of the curtain or tubing.

The following best practices will reduce dust exposure on blowing ventilation sections:

- The operator should be positioned in the mouth of the blowing line curtain with intake air sweeping from behind. The operator should not proceed past the end of the line curtain. If the operator must be on the return side of the curtain, some of the intake air should be bled over the line brattice to provide fresh air to the operator (see Figure 4-12). Good communication with shuttle car operators is essential because visibility can be a problem depending on where the continuous miner operator is standing.

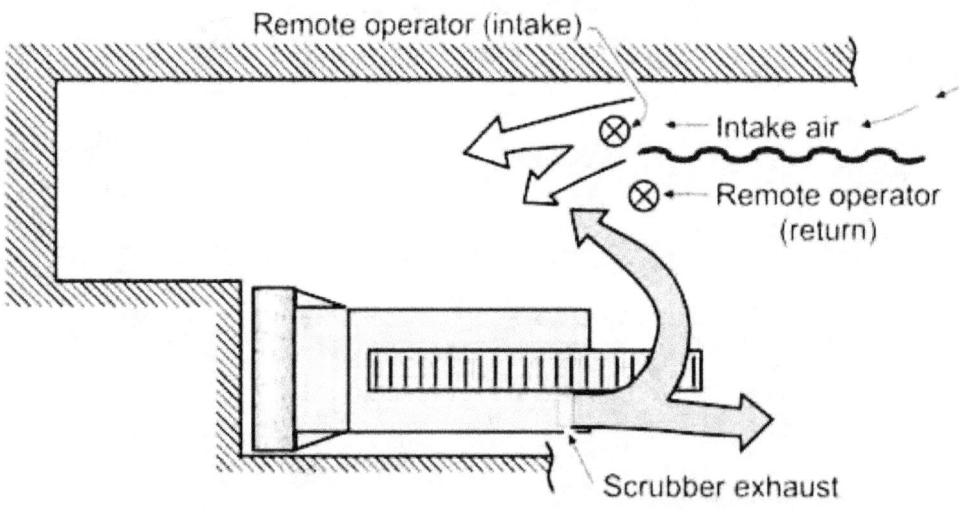

Figure 4-12.—Schematic of a blowing ventilation system.

- According to MSHA, when it is necessary for the operator to move from the clean air position (end of the curtain), the operator should allow the dust-laden air to clear the entry and stop the scrubber before moving.

- According to MSHA, when aligning the continuous miner to square a face, the operator should position the machine and then return to the end of the curtain before coal cutting resumes. This reduces the potential for injury.

- Brattice discharge velocities exceeding 800 fpm have better penetration to the face and thus better dilution of dust and methane. When brattice discharge velocities are less than 400 fpm, there is little difference in performance between blowing and exhausting ventilation [Luxner 1969].

- Scrubber discharge must be on the opposite side of the line brattice to allow scrubber exhaust to discharge directly into return air.

- The air quantity provided by the line curtain should be limited to 1,000 cfm over the scrubber capacity. Air quantities exceeding 1,000 cfm over the scrubber capacity can overpower the scrubber and push dust-laden air past the scrubber inlets [Schultz and Fields 1999]. Therefore, MSHA typically requires that the airflow entering a cut be equal to or exceed the scrubber airflow by no more than 1,000 cfm and must be measured with the scrubber off.

- Excess air velocity may be reduced by flaring out the line curtain at the end to lower the velocity of the air emerging from behind it or by pulling the line curtain back slightly to prevent overpowering the scrubber [Schultz and Fields 1999].

- Experiments have shown that erecting a short line curtain during the slab cut shields the operator from the air jet emerging from a blowing duct [Jayaraman and Jankowski 1988].

Exhausting Face Ventilation

When exhausting ventilation is used, intake air is delivered to the face in the working entry. The clean air sweeps the face, and the dust-laden air is then drawn behind the return curtain or through the exhaust tubing to the return entries. This system will keep mobile equipment in fresh air and affords the continuous miner operator more freedom of movement than a blowing ventilation system. In addition, exhausting ventilation allows more visibility around the loading area so that shuttle car operators can easily determine where the continuous miner operator is located when entering the face area.

The following best practices will reduce dust exposure on exhausting ventilation sections:

- Figure 4-13 shows a schematic of an exhaust ventilation system. Exhausting airflow allows for more flexibility than blowing, giving the operator more options to avoid dusty air. However, MSHA maintains that position A (opposite side of curtain) is preferred. As always, good communication between the continuous miner operator and shuttle car operators is essential for safe positioning.

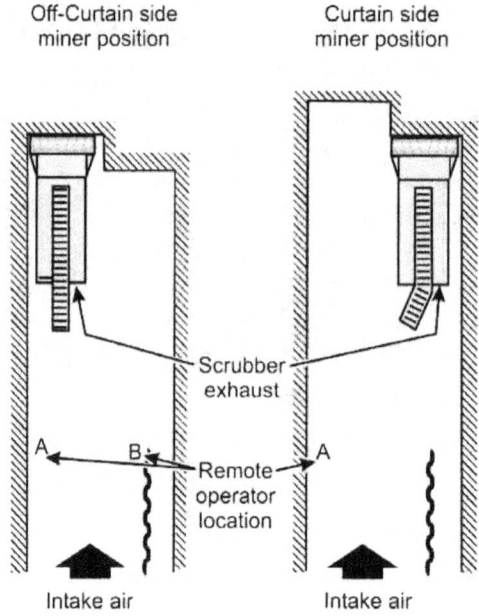

Figure 4-13.—Schematic of an exhaust ventilation system.

- An advantage of exhausting ventilation is that shuttle car operators are always positioned in fresh air.

- Air quantity reaching the inby end of the line curtain should be equal to or slightly greater than the scrubber capacity to guard against recirculation of air.

- MSHA regulations state that mean entry air velocity must be at least 60 ft/min when using exhaust ventilation systems.

- The end of the exhaust curtain or tubing must be kept within 10 ft of the face when not using a scrubber to ensure that air reaches and effectively sweeps the face.

- The operator should not proceed inby the end of the line curtain since this will expose the operator to dust-laden return air. If operator dust levels are too high, the first thing to check is whether the operator is standing parallel to or outby the end of the line curtain [Kissell and Goodman 2003].

- Scrubber exhaust must be on the same side of the entry as the line curtain to allow scrubber exhaust to discharge directly into return air [Colinet and Jankowski 1996].

DUST CONTROL FOR ROOF BOLTERS

Most roof bolting machines are equipped with MSHA-approved dry dust collection systems to remove dust during drilling. Roof bolter operators can be overexposed to dust from the drilling process, cleaning the dust collector, poor dust collector maintenance, or working downwind of the continuous mining machine. The largest source of operator dust exposure can occur from working downwind of the continuous miner. If the dry dust collector is properly maintained and if the bolter is not working downwind of the continuous miner, very little dust should be measured in the roof bolter's work environment [USBM 1984].

Three major bolter problem areas are (1) filter problems (leaks or plugging), (2) accumulation of dust in the collection system, and (3) low airflow at the bit due to hose, fitting, and relief valve leaks.

The following best practices can help reduce dust exposure to the bolter operator:

- **Maintaining the dust collector system.** Hoses and gaskets should be checked for leaks. Smoke tubes can be used to show where leaks occur. Checks should also be made for loose hose connections and damaged compartment door gaskets. Vacuum pressure at the drill head should be checked daily by using an approved pressure gauge to maintain manufacturer's vacuum specifications for proper airflow listed on the approval plate.

- **Cleaning the dust box.** Frequent cleaning of the main dust compartment is necessary to ensure proper operation of the dust collection system. The operator should maintain an upwind position when removing dust from the dust box to reduce exposure. If a rake is used to pull dust from the main compartment, care should be taken so that a dust cloud is not created and dust does not contaminate clothing. When emptying dust boxes on a dual-boom roof bolter, the return-side operator should empty the collector box first and then take a position on the intake side of the entry until the other box is emptied. Again, when performing this function, an upwind position is crucial to minimize dust exposure, and use of a respirator is recommended. Similarly, cleaning should take place in a well-ventilated entry so that liberated dust is quickly removed from the operator's breathing zone. The procedure for handling of drill cuttings is detailed in the operator's manual for most roof bolting machines.

- **Using dust collector bags.** Dust collector bags can be used with dry dust collectors to greatly reduce dust exposures when cleaning the dust box. A retrofit kit installed in the collector box enables installation of the bags (Figure 4-14) for those machines equipped with precleaner cyclones. The use of dust collector bags to contain dust in the main compartment allows workers to easily remove the dust from the main compartment and deposit it against the rib. Exposure during cleaning is reduced, the cuttings are located out of the entry traffic, and the canister filter can remain in operation for a much longer time period because of reduced dust loading [NIOSH 2007].

Figure 4-14.—Dust collector box with collector bag installed.

- **Removing and replacing the canister filter.** In past practice, canister filters were removed and impacted against a surface to dislodge caked dust from the filter media. The filter was reinserted into the collector box. Unfortunately, this practice can create a dust cloud that contaminates the breathing zone of the roof bolter operators. Cleaning the filter in this manner also creates the potential for contaminating the collector's downstream discharge components (vacuum pump and muffler) with respirable dust (Figure 4-15). When these downstream components become contaminated, respirable dust is discharged back into the mine environment in the collector's exhaust. To rectify this hazardous condition, the downstream components must be removed and cleaned as described in the next section. Although the cleaning and reuse of canister filters is still occurring, NIOSH, MSHA, and bolter manufacturers recommend that contaminated filters be replaced to minimize worker dust exposures. Replacement of filter canisters should be completed in well-ventilated entries.

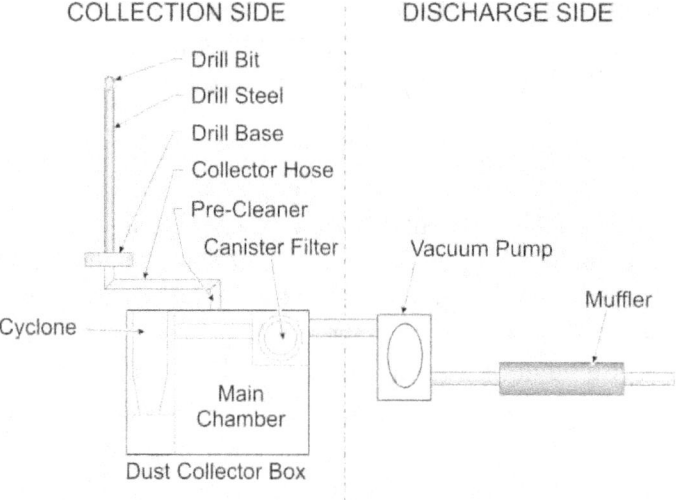

Figure 4-15.—Schematic of roof bolter dust collector components.

- **Cleaning the discharge side of the collector.** If the discharge side of the collector system becomes contaminated due to collector filter damage and/or leaks, all components downstream of the collector box must be removed and flushed with water. Surveys have shown that removing and cleaning contaminated components downstream of the canister filter results in major improvements in dust and silica levels emitted from the collector's discharge [Thaxton 1984].

- **Installing a sock on precleaners.** Some roof bolting machines are equipped with dust collector precleaners, which are cyclones collectors designed to remove larger cuttings from the airstream before reaching the collector box. These cuttings are routinely dumped after each hole is drilled and may contain some respirable dust. When the precleaner door is actuated, the dust falls an unconfined distance from the precleaner's discharge chute to the mine floor. Some dust can be entrained into the ventilating air as it falls to the mine floor. This entrained dust can be minimized by attaching a "sock" made from brattice or rubber belting as an extension to the chute to minimize the distance the dust falls unconfined.

- **Using "dust hog" bits.** Dust hog bits have a collection port in the bit body and are more effective at capturing drill dust than shank bits, where the collection port is in the drill steel. In one study, shank bits allowed from 3 to 10 times more dust to escape from the drill hole collar than "dust hog" bits [USBM 1985]. Shank-type bits should be avoided where possible, and dull bits should be replaced immediately.

- **Positioning to avoid working downwind of the continuous miner.** Regardless of the type of ventilation being used, the cutting sequence must be designed to limit the amount of time the bolter works downwind of the continuous miner. Properly sequenced cuts with double-split ventilation can eliminate the need to work downwind of dust concentrations created by the continuous miner. However, double-split ventilation is often used on super sections (i.e., two miners and two bolters in each split). If super sections are in use, then proper cut sequencing in each split is required to minimize the hazard of working downwind of the miner.

- **Wet drilling/mist drilling.** Although wet drilling can effectively control dust emissions, this option can create difficult working conditions for operators [Kissell and Goodman 2003]. Successful dust control with wet drilling typically requires 2 gpm to be supplied to the drill hole. Water is pumped to the drilling interface through the drill steel, captures dust, and then flows out of the drill hole onto the mine floor. This water can create wet floor problems and an uncomfortable work environment for the bolter operators. Mist drilling attempts to reduce the water application while still maintaining effective dust control. Reduced quantities of water and compressed air are supplied to the drilling interface in an effort to capture dust. Mist drilling typically uses less than 0.5 gpm. Although more desirable from an operations perspective, this method has yet to be shown to be as effective as wet drilling or properly operating dry collection systems [Beck and Goodman 2008].

- **Canopy air curtain.** In addition to dust created by the roof bolter itself, bolter operators can also be exposed to miner-created dust when bolting is required downwind of the continuous miner. NIOSH is currently developing and testing a device to deliver fresh air over the operator's breathing area that would reduce dust exposure while working downwind of the continuous miner [Goodman and Organiscak 2002]. A fan located at the rear of the bolter draws in entry air, filters the air, and supplies it to the area beneath the canopy through tubing and a plenum (Figure 4-16). The plenum is the same shape as the canopy and covers the same area so that protection is provided while the operator is under any portion of the canopy. Laboratory tests have shown that this concept reduced dust levels under the canopy by nearly 60%.

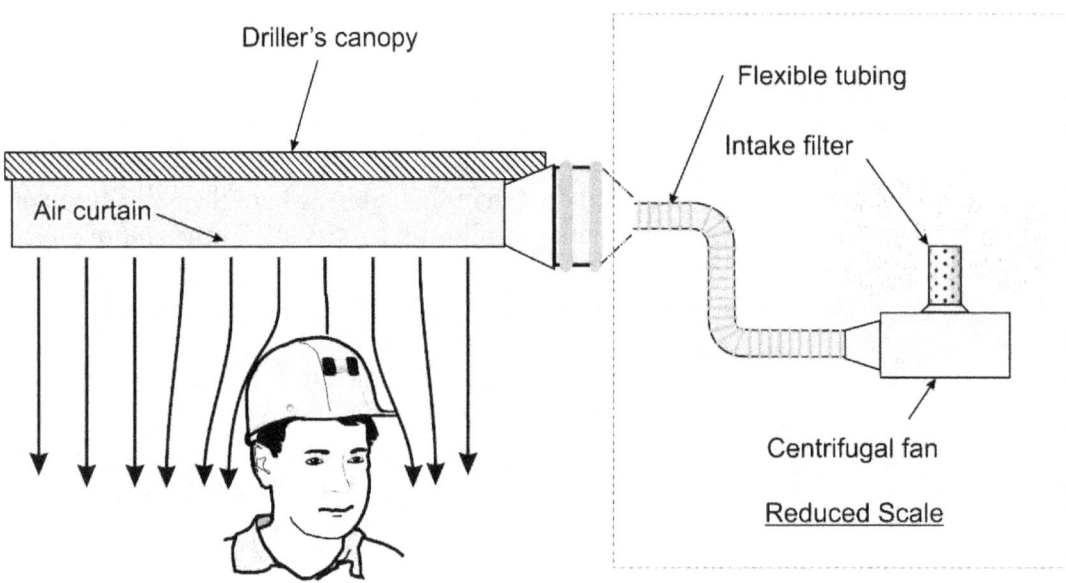

Figure 4-16.—Prototype of canopy curtain.

- **Routing miner-generated dust to the return.** For controlling respirable dust exposure downwind of the continuous miner on sections that ventilate with exhausting auxiliary fans, the simplest and most effective technique is to route the miner-generated dust directly to the return using lightweight, collapsible tubing [Jayaraman et al. 1989]. This allows the dust to reach the return without exposing downwind workers.

- **Working downwind of the bolter.** When downwind of the bolter, as much as 25% of the continuous miner operator's quartz dust exposure can be attributed to dust from the bolting operation. The problem is usually a lack of maintenance on the bolter dust collector [Organiscak et al. 1990].

INTAKE AIRWAYS

The average concentration of respirable dust in intake air must be kept at or below 1 mg/m^3 within 200 ft outby the working face. However, to maintain consistent dust control in the face area, MSHA recommends that intake concentrations be less than 0.5 mg/m^3 [Shultz and Fields 1999]. Maintaining this concentration is not usually difficult, but requires attention from mine operators to address activities that can raise intake air dust levels. Typically, high levels of intake dust are sporadic and brief in nature due to activities in the intake entries that may take place over the course of a working shift. These sporadic activities include:

- Delivery of supplies and/or personnel
- Parking equipment in intake
- Rock dusting
- Scoop activity
- Construction activity

In addition, the belt entry can be used to bring intake air to the working faces and is a potential source of dust generation. If intake dust levels are high, the following steps can be taken to maintain dust levels to a minimum:

- Good housekeeping practices will help keep intake entries free of debris, equipment, and supplies.

- Supply delivery, scoop activity, stopping construction, and rock dusting should be dedicated to nonproduction shifts.

- If haulage activities must take place during a production shift, the haulage roadways should be kept damp at all times. Since water will likely evaporate in the ventilation air, a hygroscopic salt or effective dust-allaying agent should be used [Ondrey et al. 1994]. Keeping dust dampened in the main intake entries will limit dust entrained by activity in these entries.

- Equipment should be parked in crosscuts to keep main airways clear of obstruction.

When belt air is used for face ventilation, dust generated in the belt area should be controlled. Potts and Jankowski [1992] measured the dust level impact of using belt air for face ventilation on continuous miner sections. Controls at the belt head helped maintain low dust levels in the belt entry. Automated sprays were used to suppress dust at the section-to-main transfer point. A belt scraper equipped with sprays controlled dust by cleaning the outside surface of the belt after the coal had been transferred to the main belt. These measures are also discussed in more detail in Chapter 3.

FEEDER-BREAKERS AND SHUTTLE CARS

Dust measurements show that feeder-breaker operations can contribute a significant amount of respirable dust to belt entry air, which emphasizes the need for dust controls at this location [Potts and Jankowski 1992]. Outby areas can be placed on a more stringent dust standard due to the presence of respirable silica dust. A study by Organiscak et al. [1990] showed elevated respirable silica dust concentrations at the feeder-breaker and on the shuttle car. Following are some basic controls for these areas:

- MSHA recommends hollow- or full-cone sprays at the feeder-breaker transfer point to wet and knock down coal and silica dust [Ondrey et al. 1994].

- When shuttle cars unload, dust levels can be decreased by using automated sprays at the mouth of the feeder-breaker that activate during dumping to wet coal before it enters the crusher.

- Throat sprays on the continuous miner will wet coal when entering the conveyor and lessen dust when transferred to and from shuttle cars. Redistributing a small portion of the water available on the continuous mining machine to the chain conveyor may be necessary to ensure that the loaded coal is wet enough to minimize dust reentrainment at the section loading point [Ondrey et al. 1994].

- Shuttle cars should not be in a waiting position beneath check curtains.

- Shuttle car operators should not be located in the direct discharge of the dust collector (scrubber) on the continuous miner.

- When blowing ventilation is used, configure shuttle car runs to minimize the amount of time spent in return air.

REFERENCES

Beck TW, Goodman GVR [2008]. Evaluation of dust exposures associated with mist drilling technology for roof bolters. Min Eng *60*(12):35–39.

CFR. Code of federal regulations. Washington, DC: U.S. Government Printing Office, Office of the Federal Register.

Colinet JF, Jankowski RA [1996]. Dust control considerations for deep-cut faces when using exhaust ventilation and a flooded-bed scrubber. In: Transactions of Society for Mining, Metallurgy, and Exploration, Inc. Vol. 302. Littleton, CO: Society for Mining, Metallurgy, and Exploration, Inc., pp. 104–111.

Colinet JF, Jankowski RA [2000]. Silica collection concerns when using flooded-bed scrubbers. Min Eng *52*(4):49–54.

Divers EF [1976]. Nonclogging water spray system for continuous mining machines: installation and operating guidelines. Pittsburgh, PA: U.S. Department of the Interior, Bureau of Mines, IC 8727. NTIS No. PB 265 934.

Fields KG, Atchison DJ, Haney RA [1991]. Evaluation of dust control for deep cut coal mining systems using a machine mounted dust collector. In: Proceedings of the Third

Symposium on Respirable Dust in the Mineral Industries. Littleton, CO: Society for Mining, Metallurgy, and Exploration, Inc., pp. 349–353.

Goodman GVR [2000]. Using water sprays to improve performance of a flooded-bed dust scrubber. Appl Occup Env Hyg *15*(7):550–560.

Goodman GVR, Organiscak JA [2002]. An evaluation of methods for controlling silica dust exposures on roof bolters. SME preprint 02-163. Littleton, CO: Society for Mining, Metallurgy, and Exploration, Inc.

Goodman GVR, Pollock DE, Beck TW [2004]. A comparison of a directional spray system and a flooded-bed scrubber for controlling respirable dust exposures and face gas concentrations. In: Ganguli R, Bandopadhyay S, eds. Mine ventilation: Proceedings of the 10th U.S./North American Mine Ventilation Symposium (Anchorage, AK, May 16–19, 2004). Leiden, Netherlands: Balkema, pp. 241–248.

Goodman GVR, Beck TW, Pollock DE, Colinet JF, Organiscak JA [2006]. Emerging technologies control respirable dust exposures for continuous mining and roof bolting personnel. In: Mutmansky JM, Ramani RV, eds. Proceedings of the 11th U.S./North American Mine Ventilation Symposium (University Park, PA, June 5–7, 2006). London: Taylor & Francis Group, pp. 211–216.

Hirschi JC, Chugh YP, Saha A, Mohany M [2002]. Evaluating the use of surfactants to enhance dust control efficiency of wet scrubbers for Illinois coal seams. In: De Souza E, ed. Proceedings of the North American/Ninth U.S. Mine Ventilation Symposium (Kingston, Ontario, Canada). Lisse, Netherlands: Balkema, pp. 601–606.

Jayaraman NI, Kissell FN, Schroeder W [1984]. Modify spray heads to reduce dust rollback on miners. Coal Age *89*(6):56–57.

Jayaraman NI, Jankowski RA [1988]. Atomization of water sprays for quartz dust control. Appl Ind Hyg *3*(12):327–331.

Jayaraman NI, McClelland JJ, Jankowski RA [1988]. Reducing quartz dust with flooded-bed scrubber systems on continuous miners. In: Proceedings of the Seventh International Pneumoconiosis Conference (Pittsburgh, PA), pp. 86–93.

Jayaraman NI, Babbitt CA, O'Green J [1989]. Ventilation and dust control techniques for personnel downwind of continuous miner. In: Transactions of Society for Mining, Metallurgy, and Exploration, Inc. Vol. 284. Littleton, CO: Society for Mining, Metallurgy, and Exploration, Inc., pp. 1823–1826.

Jayaraman NI, Colinet JF, Jankowski RA [1992]. Recent Basic research on dust removal for coal mine applications. In: Proceedings of the Fifth International Mine Ventilation Congress (Johannesburg, Republic of South Africa), pp. 395–405.

Kissell FN [2003]. Dust control methods in tunnels and underground mines. In: Kissell FN, ed. Handbook for dust control in mining. Pittsburgh, PA: U.S. Department of Health and Human Services, Centers for Disease Control and Prevention, National Institute for Occupational Safety and Health, DHHS (NIOSH) Publication No. 2003-147, IC 9465, pp. 3–21.

Kissell FN, Goodman GVR [2003]. Continuous miner and roof bolter dust control. In: Kissell FN, ed. Handbook for dust control in mining. Pittsburgh, PA: U.S. Department of Health and Human Services, Centers for Disease Control and Prevention, National Institute for Occupational Safety and Health, DHHS (NIOSH) Publication No. 2003-147, IC 9465, pp. 23–38.

Luxner JV [1969]. Face ventilation in underground bituminous coal mines. Airflow and methane distribution patterns in immediate face area: line brattice. Washington, DC: U.S. Department of the Interior, U.S. Bureau of Mines, RI 7223.

MSHA [2009]. Standardized Information System: Respirable coal mine quartz dust data. Arlington, VA: U.S. Department of Labor, Mine Safety and Health Administration.

NIOSH [1997]. Hazard identification 1: Exposure to silica dust on continuous mining operations using flooded-bed scrubbers. Cincinnati, OH: U.S. Department of Health and Human Services, Centers for Disease Control and Prevention, National Institute for Occupational Safety and Health, HID1, DHHS (NIOSH) Publication No. 97–147.

NIOSH [2007]. Technology news 523: Evaluation of dust collector bags for reducing dust exposure of roof bolter operators. Pittsburgh, PA: U.S. Department of Health and Human Services, Centers for Disease Control and Prevention, National Institute for Occupational Safety and Health, DHHS (NIOSH) Publication No. 2007–119.

Ondrey RS, Haney RA, Tomb TF [1994]. Summary of minimum dust control parameters. In: Proceedings of the Fourth Symposium on Respirable Dust in the Mineral Industries (Pittsburgh, PA, November 8–10, 1994).

Organiscak JA, Page SJ, Jankowski RA [1990]. Sources and characteristics of quartz dust in coal mines. Pittsburgh, PA: U.S. Department of the Interior, Bureau of Mines, IC 9271. NTIS No. PB 91-160911/AS.

Organiscak JA, Khair AW, Ahmad M [1996]. Studies of bit wear and respirable dust generation. In: Transactions of Society for Mining, Metallurgy, and Exploration, Inc. Vol. 298. Littleton, CO: Society for Mining, Metallurgy, and Exploration, Inc., pp. 1932–1935.

Pollock DE, Organiscak JA [2007]. Airborne dust capture and induced airflow of various spray nozzle designs. Aerosol Sci Technol *41*(7):711–720.

Potts JD, Jankowski RA [1992]. Dust considerations when using belt entry air to ventilate work areas. Pittsburgh, PA: U.S. Department of the Interior, Bureau of Mines, RI 9426.

Schroeder WE, Babbitt C, Muldoon TL [1986]. Development of optimal water spray systems for dust control in underground mines. Foster-Miller, Inc. U.S. Bureau of Mines contract H0199070. NTIS No. PB 87-141537.

Schultz MJ, Fields KG [1999]. Dust control considerations for deep cut mining sections. SME preprint 99-163. Littleton, CO: Society for Mining, Metallurgy, and Exploration, Inc.

Strebig KC [1975]. "Wet-head" tests on miners concluded. Coal Min & Process *12*(4): 78–80, 88.

Thaxton RA [1984]. Maintenance of a roof bolter dust collector as a means to control quartz. In: Proceedings of the Coal Mine Dust Conference (Morgantown, WV, October 8–10, 1984), pp. 137–143.

USBM [1984]. Technology news 198: Better roof bolter dust collector maintenance reduces silica dust levels. Pittsburgh, PA: U.S. Department of the Interior, Bureau of Mines.

USBM [1985]. Technology news 219: Reducing dust exposure of roof bolter operators. Pittsburgh, PA: U.S. Department of the Interior, Bureau of Mines.

CHAPTER 5.—CONTROLLING RESPIRABLE SILICA DUST AT SURFACE MINES

By John A. Organiscak[1]

Overexposure to airborne respirable crystalline silica dust (referred to here as "silica dust") can cause silicosis, a serious and potentially fatal lung disease. Mining continues to have some of the highest incidences of worker-related silicosis, and the mining machine operator is the occupation most commonly associated with the disease [NIOSH 2003]. In particular, some of the most severe cases of silicosis have been observed in surface mine rock drillers [NIOSH 1992]. A voluntary surface coal miner lung screening study conducted in Pennsylvania in 1996 found that silicosis was directly related to age and years of drilling experience [CDC 2000].

U.S. mine workers continue to be at risk of exposure to excessive levels of silica dust. The percentage of Mine Safety and Health Administration (MSHA) dust samples during 2004–2008 that exceeded the applicable or reduced respirable dust standard because of the presence of silica were: 12% for sand and gravel mines, 13% for stone mines, 18% for nonmetal mines, 21% for metal operations, and 11% for coal mines [MSHA 2009]. At surface mining operations, occupations most frequently exceeding the applicable respirable dust standard are usually operators of mechanized equipment such as drills, bulldozers, scrapers, front-end loaders, haul trucks, and crushers.

This chapter summarizes the current state of the art of dust controls for surface mines. Surface mining operations present dynamic and highly variable silica dust sources. Most of the dust generated at surface mines is produced by mobile earth-moving equipment such as drills, bulldozers, trucks, and front-end loaders excavating silica-bearing rock and minerals. Four practical areas of engineering controls to mitigate surface mine worker exposure to all airborne dusts, including silica, are drill dust collection systems, enclosed cab filtration systems, controlling dust on unpaved haulage roads, and controlling dust at the primary hopper dump.

Many surface mine dust control problems can be visually observed and diagnosed. Visible airborne dust emissions generated from a particular surface mine process usually indicate that respirable silica dust can be present and potentially become a worker exposure problem. Visual dust emissions affecting nearby workers indicates that an engineering control is needed or an existing control needs maintenance. Investigating possible causes of visual dust emissions when using an engineering control often can uncover the reason for its poor dust control effectiveness. Frequent visual inspections of engineering control systems can identify needed maintenance to optimize their dust control effectiveness. Area dust sampling should be conducted, in conjunction with personal sampling, to quantify potential dust sources and examine their contribution to the worker dust exposure problem.

---

[1]Mining engineer, Office of Mine Safety and Health Research, National Institute for Occupational Safety and Health, Pittsburgh, PA.

DRILL DUST COLLECTION SYSTEMS

Drill dust is generated by compressed air (bailing airflow) flushing the drill cuttings from the hole. Dry or water-based dust collection systems are available for controlling this drill dust. Dry dust collection systems are the most common type of dust control incorporated into the drilling machine by original equipment manufacturers because of their ability to be operated in freezing temperatures. A typical dry dust collection system is shown in Figure 5-1. It is composed of a self-cleaning (compressed air back-pulsing of filters) dry dust collector sucking the dusty air from underneath the shrouded drill deck located over the hole. Ninety percent of dust emissions with this type of system are attributed to drill deck shroud leakage, drill stem bushing leakage, and dust collector dump discharge. Wet suppression is another drill dust collection method and involves injecting water into the bailing airflow traveling down the drill stem. The process of the bailing airflow, water droplets, and cuttings mixing together captures the airborne dust as they travel back up the hole. However, wet suppression is infrequently used because of operational problems in cold climates, lack of a readily accessible water supply, and shorter bit life. Studies by the U.S. Bureau of Mines and NIOSH have shown the practical aspects of optimizing these dust collection systems. These are discussed below for each dust collection method.

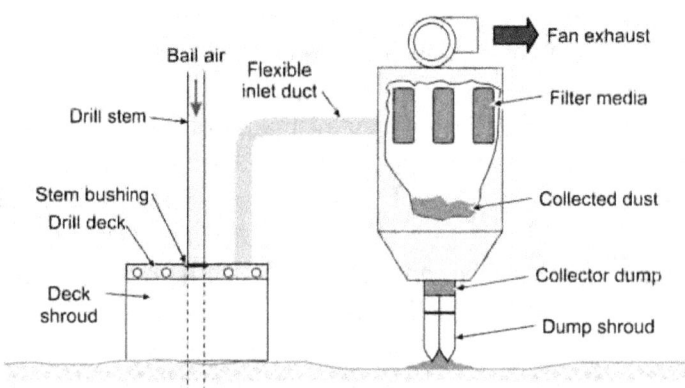

Figure 5-1.—Typical dry dust collection system used on surface drills.

Dry Dust Collector System

- **Maintain a tight drill deck shroud enclosure with the ground.** Dust emissions are significantly reduced around the drill deck shroud by maintaining the ground-to-shroud gap height below 8 in [NIOSH 2005; USBM 1987]. This can be accomplished by better vertical positioning of the drill table shroud by the operator to minimize the ground-to-shroud gap. Dust levels were significantly reduced from 21.4 to 2.5 mg/m^3 next to the drill deck shroud when the drill operator changed the drill setup procedure to minimize this gap [Organiscak and Page 1999]. Also, the ground-to-shroud gap can be more tightly closed by using a flexible shroud design that can be mechanically raised and lowered to the ground via cables and hydraulic actuators. An adjustable height shroud design maintains a better seal with uneven ground and was found to keep dust emissions next to the shroud below 0.5 mg/m^3 at several drill operations [NIOSH 1998, 2005]. Finally, a shroud constructed in sections with vertical gaps along sections or corners can also be a source of shroud leakage. Overlapping sections of shroud material reduce gaps and leakage. One conceptual shroud design for a rectangular drill table is construction with corner sections and overlapping side sections of shroud material [Page and Organiscak 1995].

- **Maintain a collector-to-bailing airflow ratio of at least 3:1.** Dust emissions are significantly decreased around the shroud at or above a 3:1 collector-to-bailing airflow ratio [NIOSH 2005]. Dust collector airflow reductions under the shroud are generally caused by restrictions and/or leakages in the system. Loaded filters and material in the ductwork are likely causes of restrictions, whereas damaged ductwork and holes are likely causes of leakage in the system. Thus, inspection and maintenance of the dust collection system are vital to achieving and maintaining optimal collector operation and airflow.

- **Maintain a good drill stem seal with the drill table.** A rubber drill stem bushing (see Figure 5-1) restricts bailing airflow from blowing dust and cuttings through the drill deck and therefore needs to be replaced after mechanical wear. An alternative sealing method involves using a nonmechanical compressed air ring seal manifold under the drill deck. This manifold consists of a donut-shaped pipe with closely spaced holes on the inside perimeter that discharges air jets in a radial pattern at the drill stem. The high-velocity air jets block the gap between the drill stem and deck, reducing respirable dust leakage through the drill deck by 41%–70% [Page 1991].

- **Shroud the collector dump discharge close to the ground.** Dumping dust from the collector discharge several feet above ground level can disperse significant amounts of airborne respirable dust. Dust emission reductions of greater than 63% were measured by the collector discharge dump after installing an extended shroud near ground level (Figure 5-1) [Reed et al. 2004; USBM 1995]. These shrouds can be fabricated quickly by wrapping brattice cloth around the perimeter of the collector discharge dump and securing it to the discharge dump with a hose clamp.

- **Maintain the dust collector as specified by manufacturer.** Collector system components should be frequently inspected and damaged components repaired or replaced. A 51% dust emission reduction was measured at one drill after a broken collector fan belt was replaced, while another drill showed a reduction of 83% after the torn deck shroud was replaced [Organiscak and Page 1999].

Wet Suppression

- **Add small amounts of water into the bailing air until the visible dust cloud has been significantly reduced.** Drill dust emissions are significantly reduced by increasing the water flow rate from 0.2 to 0.6 gal/min [USBM 1987]. A needle valve and water flow meter installed on the water supply line provides adjustable control for wet suppression systems. However, adding excessive water down the hole can cause operational problems with no appreciable improvement in dust control.

- **Minimizing water flow to a rolling cutter bit can increase bit life.** Wet drilling with rolling cutter bits can cause premature bit wear. A drill stem water separator installed upstream of a rolling cutter bit can increase bit life without adverse affects on dust control [Listak and Reed 2007; USBM 1988]. The water separator is a bit stabilizer with an internal cyclonic or impaction water droplet classifier, removing most of the water from the bailing airflow before it is flushed through the drill bit. The water removed by the internal separator is released through external holes in the bit stabilizer (Figure 5-2).

Figure 5-2.—Water separator discharging water before it reaches the drill bit.

ENCLOSED CAB FILTRATION SYSTEMS

Enclosed cab filtration systems are one of the mainstay engineering controls for reducing mobile equipment operators' exposure to airborne dust at surface mines. Enclosed cabs with heating, ventilation, and air conditioning (HVAC) systems are typically integrated into the drills and mobile equipment to protect the operator from the outside environment. Air filtration is often part of the HVAC system as an engineering control for airborne dusts. Surface mining dust surveys conducted by NIOSH on drills and bulldozers have shown that enclosed cabs can effectively control the operator's dust exposure, but cab performance can vary [Organiscak and Page 1999]. The enclosed cab protection factors (outside ÷ inside dust concentrations) measured on rotary drills ranged from 2.5 to 84, and those measured on bulldozers ranged from 0 to 45. NIOSH also conducted field studies of upgrading older equipment cabs to improve their dust control effectiveness. These studies involved retrofitting older enclosed cabs with air-conditioning, heating, and air filtration systems to show the effectiveness of upgrading older mine equipment cabs. During these retrofits, any reasonably repairable cracks, gaps, or openings were sealed with silicone and closed cell foam tape. Varying degrees of enclosure integrity were achieved. Table 5-1 shows the results in ascending order of performance achieved with these retrofitted installations. In addition, NIOSH conducted controlled laboratory experiments to examine the key design factors of enclosed cab dust filtration systems. The key performance factors for effective enclosed cab dust filtration systems are summarized below.

Table 5-1.—Respirable dust sampling results of enclosed cab field studies

Cab evaluation	Reference	Cab pressure, in w.g.	Wind velocity equivalent,[1] mph	Average inside cab dust level, mg/m^3	Average outside cab dust level, mg/m^3	Protection factor, out / in
Rotary drill	Organiscak et al. [2003a]	ND	0	0.08	0.22	2.8
Haul truck	Chekan and Colinet [2003]	0.01	4.5	0.32	1.01	3.2
Front-end loader	Organiscak et al. [2003a]	0.015	5.6	0.03	0.30	10.0
Rotary drill	Cecala et al. [2003]	0.20–0.40	20.3–28.7	0.05	2.80	56.0
Rotary drill	Cecala et al. [2005]	0.07–0.12	12.0–15.7	0.07	6.25	89.3

ND   None detected.
[1]Wind velocity equivalent = (4000 $\sqrt{\Delta p_{cab}}$) fpm × 0.11364 mph/fpm @ standard air temperature and pressure.

Key Performance Factors for Enclosed Cab Filtration Systems

- **Ensure good cab enclosure integrity to achieve positive pressurization against wind penetration into the enclosure.** As shown in Table 5-1, significant improvements in cab protection factors were achieved in the field studies when cab pressures exceeded 0.01 in w.g. This corresponded to wind velocity equivalents (an indicator of cab wind velocity resistance) greater than 4.5 mph. The cab enclosures with greater than 0.01 in w.g. pressure were of close-fitted construction, and their integrity could be readily improved by sealing cab enclosure cracks, gaps, or openings with silicone and closed cell foam tape. The loosely fitted cab construction on one of the drills and the truck were difficult to seal, which limited the amount of cab pressure that could be attained.

- **Use high-efficiency respirable dust filters on the intake air supply into the cab.** Filter efficiency performance specifications used in the field were 95% or greater on respirable-sized dusts [Chekan and Colinet 2003; Cecala et al. 2003, 2005; Organiscak et al. 2003a]. Laboratory experiments showed an order of magnitude increase in cab protection factors when using a 99% efficient filter versus a 38% efficient filter on respirable-sized particles [NIOSH 2007].

- **Use an efficient respirable dust recirculation filter.** All of the cab field demonstrations used recirculation filters that were 95% efficient or better in removing respirable-sized dusts [Chekan and Colinet 2003; Cecala et al. 2003, 2005; Organiscak et al. 2003a]. Laboratory experiments showed an order of magnitude increase in cab protection factors when using an 85%–94.9% efficient filter compared to no recirculation filter [NIOSH 2007]. Laboratory testing also showed that when using a recirculation filter, the time for interior cab concentration to decrease and reach stability after the door had been opened and closed was cut by more than half.

- **Minimize dust sources in the cab.** Use good housekeeping practices, and move heater outlets that blow across soiled cab floors. Dust levels were shown to increase from 0.03 to 0.26 mg/m^3 by turning on a floor heater inside the cab [Cecala et al. 2005]. The floor heater was removed and cab heating was discharged down from the ceiling HVAC system, reducing dust entrainment in the cab during colder winter months. Another method of reducing entrainment of dust from a soiled cab floor is placing a gritless (without sand added) sweeping compound on the floor during the working shift. Most commercial sweeping compounds have petroleum-based oils or wax added to the cellulose material. It must be noted, however, that people sensitized to petroleum distillates could have allergic reactions to these sweeping compounds if used in enclosed cabs. A few companies offer non-petroleum-based sweeping compounds that use either a natural oil or chemical additive for dust adhesion [NIOSH 2001]. It is also recommended to cover the floor with rubber matting instead of carpeting for easy cleaning. More frequent cleaning of heavily soiled floors by the operator may be a more straightforward alternative to using sweeping compounds to minimize this type of dust entrainment.

- **Keep doors closed during equipment operation.** On one drill operation, the respirable dust concentrations inside the cab averaged 0.09 mg/m^3 with the door closed and 0.81 mg/m^3 when the door was briefly opened to add drill steels [Cecala et al. 2007]. Although this occurred after drilling stopped and the visible dust dissipated, opening the door, even briefly, produced a ninefold increase in respirable dust concentrations inside the cab during the many drill steel changes made over a working shift.

CONTROLLING HAULAGE ROAD DUST

Off-road haul trucks used in the mining industry typically contribute most of the total dust emissions at a mine site. Although most of the airborne dust generated from unpaved haulage roads is nonrespirable, up to 20% is in the respirable size range [Organiscak and Reed 2004]. The most common method of haul road dust control is surface wetting with water. Figure 5-3 shows the effectiveness of road wetting with water on airborne respirable dust generation measured next to an unpaved haul road. The road was wetted in the morning and dried out in the afternoon. Although the road treatment methods have been shown to be very effective, their application generally involves continual maintenance due to road degradation from traffic, dry climatic conditions, and material spillage on the road. Road dust generation then can be inevitable at times during the mining operation until controls are applied. Given their mobility, trucks have the potential to expose other downwind mine workers to respirable dust, as well as other truck drivers on the haul road. NIOSH has recently studied the size characteristics, concentrations, and spatial variation of airborne dust generated along unpaved mine haulage roads to examine the potential human health and safety impacts of this dust source and is examining other avenues of truck dust mitigation. Techniques for controlling haulage road dust are summarized below.

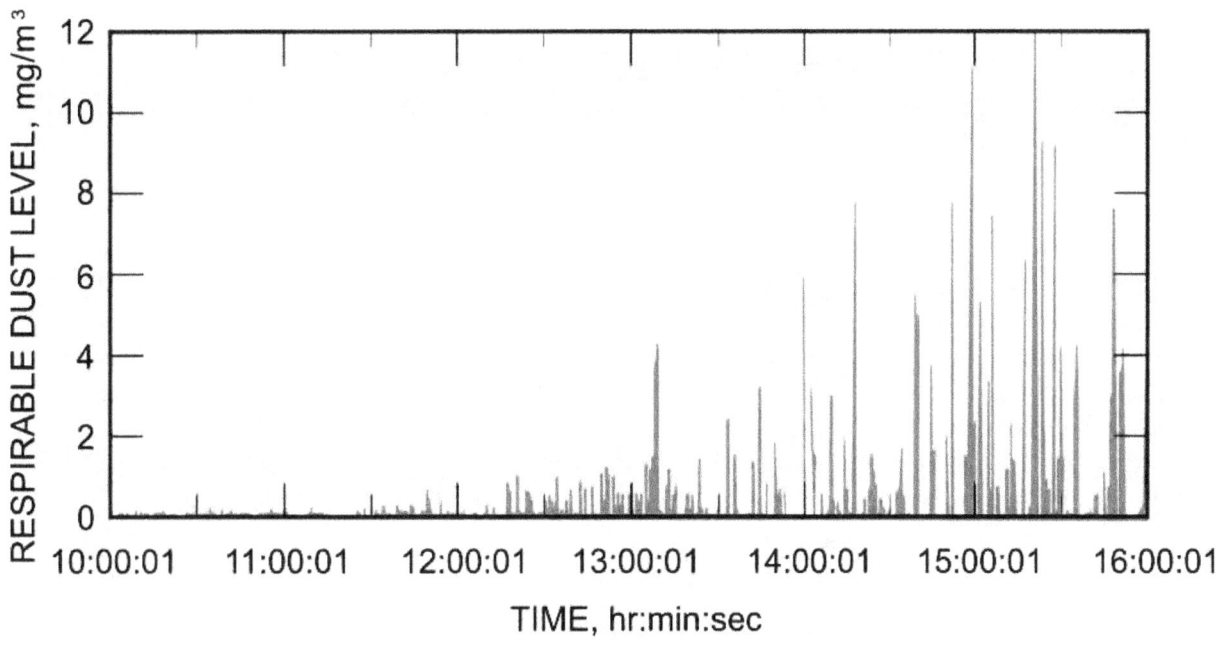

Figure 5-3.–Increase in dust when a wet haul road dries.

Methods for Controlling Haulage Road Dust Exposures

- **Treatment of unpaved road surfaces.** Figure 5-3 shows the effectiveness of road wetting with water on respirable dust liberation next to the haul road and its time-frame of effectiveness at this mine [Organiscak and Reed 2004]. Other haulage road treatments include adding hygroscopic salts, surfactants, soil cements, bitumens, and films (polymers) to the road surface, which can extend the time of effectiveness between treatments up to several weeks [Organiscak et al. 2003b; Olson and Veith 1987].

- **Increase the distance between vehicles traveling the haul road.** Research has shown that airborne dust concentrations generated from haulage roads rapidly decreased and approached ambient air dust levels 100 ft from the road [Organiscak and Reed 2004]. This road dust dissipation and dilution provides an opportunity to use administrative and mine planning controls to reduce worker dust exposure. If a trailing haul truck was not allowed to follow within 20 sec of a leading truck, the resulting distance between trucks allowed generated dust to dissipate. This led to more than a 40% reduction in respirable dust exposure to the following truck [Reed and Organiscak 2005]. Finally, advantageous road layout and traffic patterns can be designed into the mine plan to isolate the dust sources from other workers [Organiscak and Reed 2004].

CONTROLLING DUST AT THE PRIMARY HOPPER DUMP

The mined product is normally loaded into haul trucks from the surface mine pit and driven to the primary crusher location. This product is either dumped directly from the haul truck into the primary hopper feeding a crusher or dumped into a stockpile. If it is stockpiled, a front-end loader then takes the mined product and dumps it into the primary hopper. In either case during this dumping process, a dust cloud can billow out of the hopper and roll back under the truck bed or front-end loader bucket. Dust in the mined product is released from the large volume of material being dumped in a short period of time, which quickly displaces the air in the hopper and transports the airborne dust released from dumping. If the equipment operators dumping the mined product into the hopper have an effective enclosed cab filtration system (as described earlier), their exposure to this dust would be reduced. However, if other mine personnel such as crusher operators and/or maintenance workers work near this primary dump, they can be exposed to this airborne dust. Several effective control methods are available and include enclosing the hopper dump and using water sprays to suppress and contain the dust from rolling back out of the enclosure.

Key Factors for Controlling Dust From the Primary Dump

- **Enclose the primary hopper dump.** Walls can be constructed around the primary dump location to form an enclosure that must be custom-designed to accommodate the dump vehicles being used. Walls can be either stationary (rigid) or removable (flexible material or curtains) based on maintenance access within parts of the enclosure. Staging curtains, also called stilling curtains, can be used in the enclosure to provide physical barriers that break up the natural tendency for dust to billow out of the primary dump hopper when a large volume of product is dumped in a very short time period (see Figure 5-4) [Weakly 2000]. Another option to restrict the dust from escaping the enclosure is using panels of flexible plastic stripping on the dump side of the enclosure. This plastic stripping employs an overlapping sequence that provides a very effective seal and resists damage if contacted by the bucket of the front-end loader or the bed of the haul truck during dumping. Finally, a local exhaust ventilation system can be used to filter the dust-laden air from the enclosed hopper area. This would be most appropriate when the primary dump is at a location where the dust could enter an adjoining structure or impact outside miners. Since hoppers are usually large, a significant amount of airflow would be required to create a negative pressure sufficient enough to contain the dust cloud. This approach would be a more expensive alternative than using wet suppression [Rodgers et al. 1978].

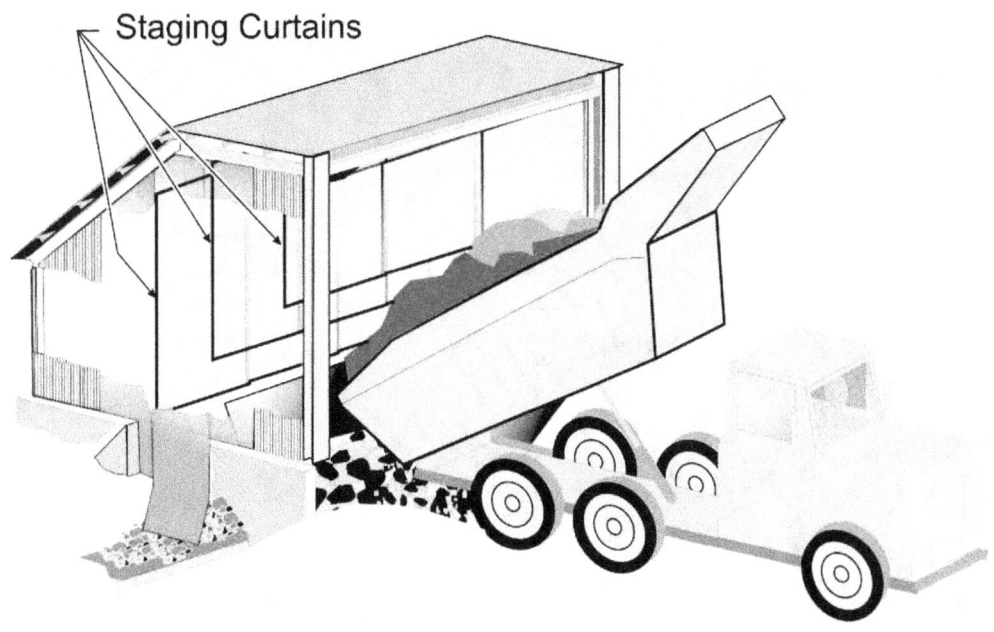

Figure 5-4.—Staging curtains used to prevent dust from billowing out of enclosure.

- **Use water sprays to suppress the dust in the enclosure.** Water sprays directed at the mined product dumped into the hopper will wet the material and suppress some of the airborne dust generated. A good starting point is to add 1% moisture by weight [Quilliam 1974]. This percentage can be adjusted based on the improvement gained from additional moisture versus any consequences from adding too much. Since continuous use of water sprays during long periods of idle time between dumping can have adverse operational effects, activate the water sprays during the actual dump cycle through the use of a photo cell or a mechanical switching device. A delay timer can also be used in this application so that the sprays continue to operate and suppress dust for a short time period after the dump vehicle has moved away.

- **Prevent the dust from rolling back under the dump vehicle.** A tire-stop water spray system is recommended to reduce the dust liberated due to rollback under the dumping mechanism. A tire stop or Jersey barrier should be positioned at the most forward point of dumping for the primary hopper. A water spray system should be attached to the back side of this tire stop to knock down and force the dust that would otherwise roll back under the dumping mechanism into the hopper. In addition, a shield should be placed over this water spray manifold to protect it from damage from falling material (Figure 5-5). Finally, a system should also be incorporated that allows the water sprays to be activated only during the actual dumping process, as previously discussed.

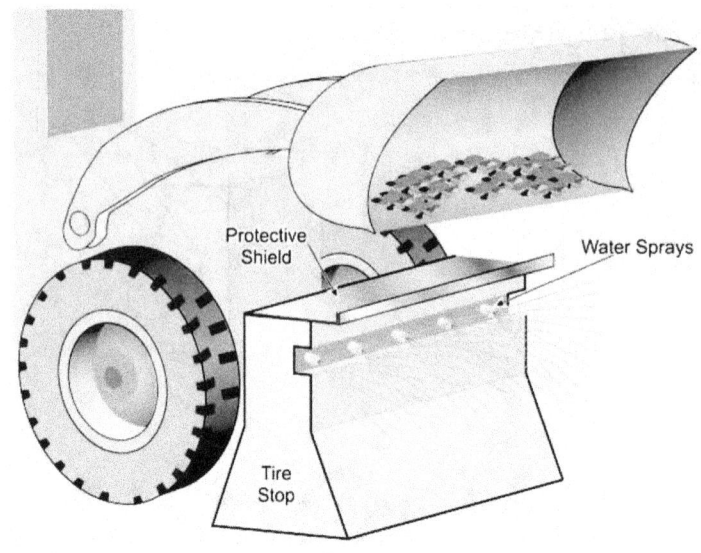

Figure 5-5.—Tire-stop water spray system reduces dust rollback under the dumping vehicle.

REFERENCES

Cecala AB, Organiscak JA, Heitbrink WA, Zimmer JA, Fisher T, Gresh RE, Ashley JD [2003]. Reducing enclosed cab drill operator's respirable dust exposure at a surface coal operation using a retrofitted filtration and pressurization system. In: Yernberg WR, ed. Transactions of Society for Mining, Metallurgy, and Exploration, Inc. Vol. 314. Littleton, CO: Society for Mining, Metallurgy, and Exploration, Inc., pp. 31–36.

Cecala AB, Organiscak JA, Zimmer JA, Heitbrink WA, Moyer ES, Schmitz M, Ahrenholtz E, Coppock CC, Andrews EH [2005]. Reducing enclosed cab drill operator's respirable dust exposure with effective filtration and pressurization techniques. J Occup Environ Hyg $2(1)$:54–63.

Cecala AB, Organiscak JA, Zimmer JA, Moredock D, Hillis M [2007]. Closing the door to dust when adding drill steels. Rock Prod $110(10)$:29–32.

CDC (Centers for Disease Control and Prevention) [2000]. Silicosis screening in surface coal miners: Pennsylvania, 1996–1997. MMWR $49(27)$:612–615.

Chekan GJ, Colinet JF [2003]. Retrofit options for better dust control. Aggregates Manag $8(9)$:9–12.

Listak JM, Reed WR [2007]. Water separator shows potential for reducing respirable dust generated on small-diameter rotary blasthole drills. Int J Min Reclam Environ $21(3)$:160–172.

MSHA [2009]. Program Evaluation and Information Resources, Standardized Information System. Arlington, VA: U.S. Department of Labor, Mine Safety and Health Administration.

NIOSH [1992]. NIOSH alert: Request for assistance in preventing silicosis and deaths in rock drillers. Cincinnati, OH: U.S. Department of Health and Human Services, Centers for Disease Control, National Institute for Occupational Safety and Health, DHHS (NIOSH) Publication No. 92–107.

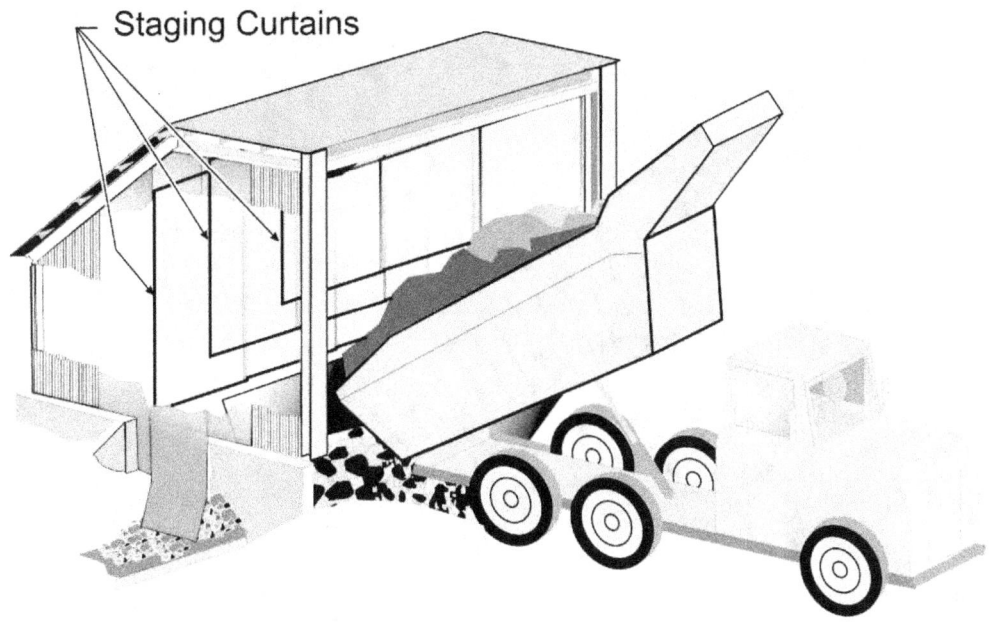

Figure 5-4.—Staging curtains used to prevent dust from billowing out of enclosure.

- **Use water sprays to suppress the dust in the enclosure.** Water sprays directed at the mined product dumped into the hopper will wet the material and suppress some of the airborne dust generated. A good starting point is to add 1% moisture by weight [Quilliam 1974]. This percentage can be adjusted based on the improvement gained from additional moisture versus any consequences from adding too much. Since continuous use of water sprays during long periods of idle time between dumping can have adverse operational effects, activate the water sprays during the actual dump cycle through the use of a photo cell or a mechanical switching device. A delay timer can also be used in this application so that the sprays continue to operate and suppress dust for a short time period after the dump vehicle has moved away.

- **Prevent the dust from rolling back under the dump vehicle.** A tire-stop water spray system is recommended to reduce the dust liberated due to rollback under the dumping mechanism. A tire stop or Jersey barrier should be positioned at the most forward point of dumping for the primary hopper. A water spray system should be attached to the back side of this tire stop to knock down and force the dust that would otherwise roll back under the dumping mechanism into the hopper. In addition, a shield should be placed over this water spray manifold to protect it from damage from falling material (Figure 5-5). Finally, a system should also be incorporated that allows the water sprays to be activated only during the actual dumping process, as previously discussed.

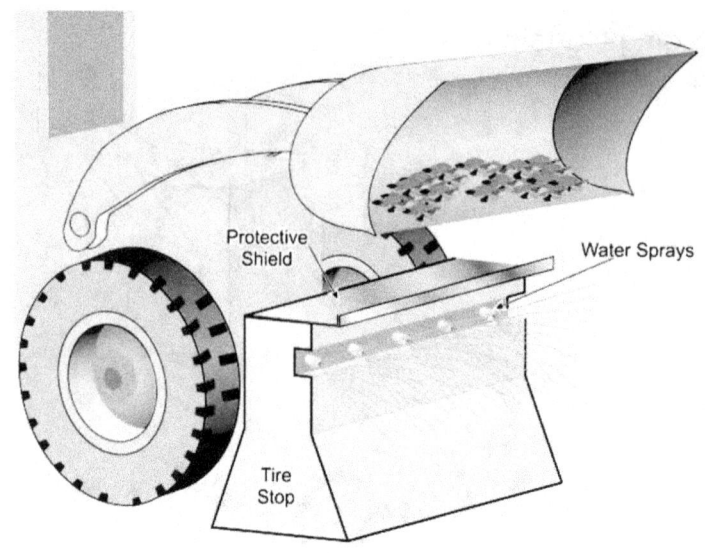

Figure 5-5.—Tire-stop water spray system reduces dust rollback under the dumping vehicle.

REFERENCES

Cecala AB, Organiscak JA, Heitbrink WA, Zimmer JA, Fisher T, Gresh RE, Ashley JD [2003]. Reducing enclosed cab drill operator's respirable dust exposure at a surface coal operation using a retrofitted filtration and pressurization system. In: Yernberg WR, ed. Transactions of Society for Mining, Metallurgy, and Exploration, Inc. Vol. 314. Littleton, CO: Society for Mining, Metallurgy, and Exploration, Inc., pp. 31–36.

Cecala AB, Organiscak JA, Zimmer JA, Heitbrink WA, Moyer ES, Schmitz M, Ahrenholtz E, Coppock CC, Andrews EH [2005]. Reducing enclosed cab drill operator's respirable dust exposure with effective filtration and pressurization techniques. J Occup Environ Hyg *2*(1):54–63.

Cecala AB, Organiscak JA, Zimmer JA, Moredock D, Hillis M [2007]. Closing the door to dust when adding drill steels. Rock Prod *110*(10):29–32.

CDC (Centers for Disease Control and Prevention) [2000]. Silicosis screening in surface coal miners: Pennsylvania, 1996–1997. MMWR *49*(27):612–615.

Chekan GJ, Colinet JF [2003]. Retrofit options for better dust control. Aggregates Manag *8*(9):9–12.

Listak JM, Reed WR [2007]. Water separator shows potential for reducing respirable dust generated on small-diameter rotary blasthole drills. Int J Min Reclam Environ *21*(3):160–172.

MSHA [2009]. Program Evaluation and Information Resources, Standardized Information System. Arlington, VA: U.S. Department of Labor, Mine Safety and Health Administration.

NIOSH [1992]. NIOSH alert: Request for assistance in preventing silicosis and deaths in rock drillers. Cincinnati, OH: U.S. Department of Health and Human Services, Centers for Disease Control, National Institute for Occupational Safety and Health, DHHS (NIOSH) Publication No. 92–107.

NIOSH [1998]. Hazard controls: New shroud design controls silica dust from surface mine and construction blast hole drills. Cincinnati, OH: U.S. Department of Health and Human Services, Centers for Disease Control and Prevention, National Institute for Occupational Safety and Health, HC27, DHHS (NIOSH) Publication No. 98–150.

NIOSH [2001]. Technology news 487: Sweeping compound application reduces dust from soiled floors within enclosed operator cabs. Pittsburgh, PA: U.S. Department of Health and Human Services, Centers for Disease Control and Prevention, National Institute for Occupational Safety and Health.

NIOSH [2003]. Work-related lung disease surveillance report, 2002. Morgantown, WV: U.S. Department of Health and Human Services, Centers for Disease Control and Prevention, National Institute for Occupational Safety and Health, DHHS (NIOSH) Publication No. 2003-111.

NIOSH [2005]. Technology news 512: Improve drill dust collector capture through better shroud and inlet configurations. Pittsburgh, PA: U.S. Department of Health and Human Services, Centers for Disease Control and Prevention, National Institute for Occupational Safety and Health, DHHS (NIOSH) Publication No. 2006–108.

NIOSH [2007]. Technology news 528: Recirculation filter is key to improving dust control in enclosed cabs. Pittsburgh, PA: U.S. Department of Health and Human Services, Centers for Disease Control and Prevention, National Institute for Occupational Safety and Health, DHHS (NIOSH) Publication No. 2008–100.

Olson KS, Veith DL [1987]. Fugitive dust control for haulage roads and tailing basins. Minneapolis, MN: U.S. Department of the Interior, Bureau of Mines, RI 9069.

Organiscak JA, Page SJ [1999]. Field assessment of control techniques and long-term dust variability for surface coal mine rock drills and bulldozers. Int J Surf Min Reclam Env *13*:165–172.

Organiscak JA, Reed WR [2004]. Characteristics of fugitive dust generated from unpaved mine haulage roads. Int J Surface Min Reclam Environ *18*(4):236–252.

Organiscak JA, Cecala AB, Thimons ED, Heitbrink WA, Schmitz M, Ahrenholtz E [2003a]. NIOSH/industry collaborative efforts show improved mining equipment cab dust protection. In: Yernberg WR, ed. Transactions of Society for Mining, Metallurgy, and Exploration, Inc. Vol. 314. Littleton, CO: Society for Mining, Metallurgy, and Exploration, Inc., pp. 145–152.

Organiscak JA, Page SJ, Cecala AB, Kissell FN [2003b]. Surface mine dust control. In: Kissell FN, ed. Handbook for dust control in mining. Pittsburgh, PA: U.S. Department of Health and Human Services, Centers for Disease Control and Prevention, National Institute for Occupational Safety and Health, DHHS (NIOSH) Publication No. 2003-147, IC 9465, pp. 73–81.

Page SJ [1991]. Respirable dust control on overburden drills at surface mines. In: Proceedings of the American Mining Congress Coal Convention, pp. 523–539.

Page SJ, Organiscak JA [1995]. Taming the dust devil: an evaluation of improved dust controls for surface drills using rotoclone collectors. Eng Min J *Nov*:30–31.

Quilliam JH [1974]. Sources and methods of control of dust. In: The ventilation of South African gold mines. Yeoville, Republic of South Africa: The Mine Ventilation Society of South Africa.

Reed WR, Organiscak JA [2005]. Evaluation of dust exposure to truck drivers following the lead haul truck. In: Yernberg WR, ed. Transactions of Society for Mining, Metallurgy, and

Explorations, Inc. Vol. 318. Littleton, CO: Society for Mining, Metallurgy, and Exploration, Inc., pp. 147–153.

Reed WR, Organiscak JA, Page SJ [2004]. New approach controls dust at the collector dump point. Eng Min J *205*(7):29–31.

Rodgers SJ, Rankin RL, Marshall MD [1978]. Improved dust control at chutes, dumps, transfer points, and crushers in noncoal mining operations. MSA Research Corp. U.S. Bureau of Mines contract No. H0230027. NTIS No. PB297-422.

USBM [1987]. Technology news 286: Optimizing dust control on surface coal mine drills. Pittsburgh, PA: U.S. Department of the Interior, Bureau of Mines.

USBM [1988]. Technology news 308: Impact of drill stem water separation on dust control for surface coal mines. Pittsburgh, PA: U.S. Department of the Interior, Bureau of Mines.

USBM [1995]. Technology news 447: Dust collector discharge shroud reduces dust exposure to drill operators at surface coal mines. Pittsburgh, PA: U.S. Department of the Interior, Bureau of Mines.

Weakly A [2000]. Controlling dust without using bag houses. Coal Age *Nov.*24–26.

www.ingramcontent.com/pod-product-compliance
Lightning Source LLC
Chambersburg PA
CBHW081840170526
45167CB00007B/2855